CIRIA C563

London, 2001

Sustainable construction:
company indicators

WS Atkins Consultants

CIRIA *sharing knowledge ■ building best practice*

6 Storey's Gate, Westminster, London SW1P 3AU
TELEPHONE 020 7222 8891 FAX 020 7222 1708
EMAIL enquiries@ciria.org.uk
WEBSITE www.ciria.org.uk

Foreword

In April 2000, the DETR published Building a *Better Quality of Life* (DETR, 2000), a document which set out a strategy for the construction industry to move towards building more sustainable assets and working in a more efficient and socially and environmentally responsible manner. The strategy was developed in close consultation with the industry and with the broad support of all of the major construction groupings.

To deliver more sustainable construction, and access the business benefits that derive from operating more sustainably, companies within the industry must strive to improve their performance. A key step to support this improvement process has been seen as the development of a framework, with series of indicators and targets, through which both individual companies and the overall industry can measure progress. This report presents outcomes of CIRIA's project RP609, undertaken with the support of leading construction companies and industry bodies and groups, to develop indicators and targets for sustainable construction.

The report describes a range of indicators to measure the sustainability of a company providing design or construction services; and to measure company progress in delivering more sustainable construction projects. Aimed primarily at managers responsible for environmental and social performance, it provides a step-by-step guide to developing and using indicators while highlighting the subsequent business benefits.

Sustainable construction: company indicators

Construction Industry Research and Information Association

C563 © CIRIA 2001 ISBN 0 86017 563 4

Keywords		
Benchmarking and KPIs, contaminated land, environmental good practice, health and safety, materials, pollution prevention, respect for people, sustainable construction, sustainable resource use, urban regeneration, waste minimisation, water quality.		
Reader interest	**Classification**	
Directors, health, safety and environmental managers	AVAILABILITY CONTENT STATUS USER	Unrestricted Guidance document Committee-guided Construction professional

Published by CIRIA, 6 Storey's Gate, Westminster, London SW1P 3AU.

*** NOTE**

...at the time of going to press Government reorganisation has meant that DETR responsibilities have been moved variously to the Department of Trade and Industry (DTI), the Department for the Environment, Food and Rural Affairs (DEFRA), and the Department for Transport, Local Government and the Regions (DTLR). References made to the DETR in this publication should be read in this context.

For clarification, readers should contact the Department of Trade and Industry.

Acknowledgements

This report is a result of the CIRIA research project RP609 *Sustainable Construction: Targets and Indicators*, which was undertaken on behalf of CIRIA by WS Atkins Consultants Ltd with the support of the New Economics Foundation. The project was funded by the DETR through the Partners in Innovation scheme, Biffaward, The Institution of Civil Engineers, and the following Construction Industry Environment Forum (CIEF) research members: Skanska Technology Ltd, Laing Technology Group Ltd, Royal Bank of Scotland Group, Ove Arup & Partners, WS Atkins Consultants Ltd.

CIRIA and the project team would like to thank members of the project steering group for their significant contributions throughout the project. The steering group comprised:

George Martin (Chair)	Forum for the Future
Peter Charnley	Royal Bank of Scotland Group
Chris Chiverrell	Laing Technology Group Ltd
Steve Creed	Stanger Science and Environment
David Crowhurst	BRE
Dave Hampton	ABS Consulting/Construction Industry Council
Paul Jackson	Skanska Technology Ltd
Tony Parry	TRL
Bruce Sharpe	DETR
Sarah Willis	Environmental Governance
Roger Berry	FBE Management Ltd
Peter Braithwaite	Ove Arup & Partners
Alex MacGillivray	New Economics Foundation

Corresponding members included:

David Anderson	Institution of Civil Engineers/Scottish Power plc
Sue Hobbs	Construction Best Practice Programme
Steve Milsom	EPSRC
Stephen Ratcliffe	Construction Confederation
Mervyn Bramley	Environment Agency
Paul Donnelly	Laing Technology Group Ltd
Clare Perkins	Ove Arup & Partners

CIRIA would also like to thank the companies who provided valuable assistance to the project by testing the indicators during their development. Particular thanks are due to the two partner companies Carillion Plc and Ove Arup, who provided extensive feedback on the testing. Companies involved included:

Balfour Beatty	Bovis Lend Lease
Buro Happold	Carillion
John Laing	Montgomery Watson
Ove Arup & Partners	Scott Wilson
Taylor Woodrow	WSP Group

CIRIA's Research Managers for the project were Jon Bootland and Martin Hunt. The project manager was Greg Archer of WS Atkins Consultants. Significant contributions to the project were made by Prof. Barrie Mould and Mr Andrew Goodwin of WS Atkins Consultants and Sanjiv Lingayah and Alison Pilling from the New Economics Foundation.

The project team would like to thank and acknowledge the assistance of all those individuals, companies and organisations that responded to the consultation document or participated in the consultation seminars, and specifically those companies and organisations which provided case-studies to illustrate the report.

Executive summary

The aims of sustainable construction are to produce structures that enhance the quality of life and protect the environment, and to do so efficiently, profitably and fairly. The current performance of the industry is not compatible with these aims; it is simply unsustainable. *Rethinking Construction* (DETR, 1998) and the Government's sustainable construction strategy (DETR 2000) emphasise the need for a culture change within the industry.

Delivering more sustainable construction requires action from all engaged in constructing and maintaining the built environment including those providing design, consulting and construction services. The business case for more sustainable construction is increasingly attractive as structural barriers to improvement are reduced or removed. The Government has signalled its intention to stimulate the required changes and is encouraging companies to respond. In the medium term companies that want to thrive must therefore innovate and deliver more sustainable built assets.

For more sustainable construction projects to be delivered, companies need to innovate and modify their methods of working and terms and conditions. The essential first step is for Directors to seize the opportunity to initiate a programme of activities to implement best practice in all aspects of sustainable construction within their businesses. Measuring company sustainability performance is a key element in such a programme.

This report identifies a series of quantitative measures (indicators) against which companies can measure the sustainability of their business (*strategic* indicators) and the activities they perform (*operational* indicators). Each indicator provides a snapshot of progress in one of the important dimensions of sustainable construction. Overall progress may be assessed by considering the full range of indicators. The indicators, and the report as a whole, have been developed following extensive consultation within the industry and with key stakeholders, with the objective of assisting providers of design and construction services to measure and improve the sustainability of their businesses. The approach has been endorsed by several industry groups and key stakeholders including the Construction Industry Board, the Sustainable Construction Focus Group, Movement for Innovation, the Sustainability Working Party, the Construction Industry Council and the Environment Agency.

The indicators are specifically designed to be relevant to business needs, and action-oriented. The principal purposes of the indicators are to promote implementation of better practice and to demonstrate the progress being made. The indicators are useful as part of company environmental and social reporting to facilitate dialogue with stakeholder groups. They also enable the company to benchmark its performance against competitors. Overall, use of and reporting on the indicators demonstrate that the company is innovative, well managed and responsive to stakeholder needs.

The proposed strategic indicators are relevant to most construction companies and measure the systems and processes in place to improve performance: for example, the use of an environmental management system. Operational indicators measure the performance of the company in delivering more sustainable construction projects, such as tonnes of waste to landfill per £ turnover. It is hoped that companies will use and report their performance against all of the strategic indicators – see the table opposite – and against selected operational indicators relevant to their business.

Table 1 *Strategic Indicators*

Environmental

- percentage turnover of company operations with a formal or independently certified Environmental Management System to ISO 14001 or equivalent
- percentage score calculated using the Business in the Environment (BiE) index of company environmental engagement
- percentage of projects for which an environmental assessment has been undertaken and proposed environmental mitigation measures implemented
- percentage of company sites operating under the Considerate Constructors or related scheme
- number of formal environmental or nuisance notices served due to construction activities
- percentage of services by value obtained from companies with a certified Environmental Management System to ISO 14001 or equivalent
- percentage of projects for which standards of environmental performance and social engagement have been formally agreed with the client
- percentage of projects for which whole-life costs and/or Life Cycle Assessment have been calculated and used in the design of the project or in the method of construction and materials used.

Social

- percentage of staff covered under the terms of an Investors in People or similar scheme
- percentage of annual staff turnover for permanent staff
- percentage of services by value obtained from companies operating an Investors in People or similar scheme
- reportable fatal and non-fatal accidents per 100 000 hours worked
- percentage of projects that include (and implement) a plan for stakeholder dialogue
- percentage of appropriate projects that include (and implement) a plan to consult with the end user
- proportion of turnover generated by projects undertaken under alliances or other forms of partnership working
- average client satisfaction using the KPI approach.

Economic

- profit before tax and interest as a percentage of sales
- profit before tax and interest per employee.

Please note that, throughout this publication, tables are colour coded as follows:

Environmental indicators:

Social indicators:

Economic indicators:

For ease of use, tables are indicated in appropriate colours on the edge of each page.

About this book

This publication is designed to provide both overall information and specific, practical guidance, the latter particularly for managers with environmental and social corporate performance responsibilities. Its principal provision is a series of performance indicators .

Section 1

...makes the general case – including the business case – for seeking to improve performance. It reviews progress so far, setting out the ten key themes which have been identified, as a result of extensive industry-wide consultation, as necessary for the delivery of sustainable construction. It describes current initiatives, the next steps which need to be taken and the role in the process of the development and use of indicators at four levels: industry, sector, company and project.

Sections 2, 3 and 4

...then present detailed information for each of three indicator groups: these measure environmental performance, social progress and economic growth and project delivery. For each, we set out the relevant key themes, followed by the strategic level indicators - those which measure overall company performance – and the operational indicators – those which, appropriately selected, can be used to measure performance on discrete projects and aggregated. Operational indicators are listed against likely priority issues for most companies, derived from the main themes.

Sustainability indicator type		Application		
			Design and consultancy companies	Construction companies
Strategic	Environmental	Table 2.1	✓	✓
	Social	Table 3.1	✓	✓
	Economic	Table 4.1	✓	✓
Operational	Environmental	Table 2.3	✓	
		Table 2.4		✓
	Social	Table 3.3	✓	✓
	Economic	Table 4.2	✓	✓

Section 5

...looks at how to go forward, using indicators to set company targets and derive direct benefit from sustainability reporting and benchmarking of company performance, while contributing to industry-wide measurement and progress. The section includes a checklist designed to help companies most effectively develop and use indicators.

Case studies

...at the end of this volume – referred to in Sections 1 and 5 – illustrate the progress made by several major players.

Appendices A1 and A2

...respectively set out and discuss the key themes and the framework of indicators at different levels.

Appendix A3

...provides a short selection compiled from DETR and other Government reports of targets in use.

Contents

List of figures, tables and case studies

FIGURES

TABLES

CASE STUDIES

1 Measuring sustainable construction

1.1 WHAT IS SUSTAINABLE CONSTRUCTION?

Sustainable construction is part of sustainable development, which is about ensuring a better quality of life for everyone, now and for generations to come (DETR, 1999). Sustainable development necessitates balancing these four elements:

- effective protection of the environment
- prudent use of natural resources
- social progress which recognises the needs of everyone
- maintenance of high and stable levels of economic growth and employment.

Sustainable construction is the application of sustainable development to the construction industry. The construction industry accounts for some 10 per cent of GDP and employs about one and a half million people (DETR, 2000). The buildings and structures the industry creates permanently alter our society and environment. Energy consumed within domestic and commercial buildings produces a third of the UK's carbon dioxide. The industry consumes six tonnes of material per year per person and produces tens of millions of tonnes of waste, a significant proportion of which is materials delivered to site and left unused. Clearly, improving the performance of the construction industry is vital to the delivery of sustainable development.

1.1.1 Building a better quality of life

Following extensive consultation with the construction industry the DETR published *Building a Better Quality of Life* (DETR, 2000) which presents a strategy for more sustainable construction. The strategy identifies these objectives for the industry:

- be more profitable and more competitive
- deliver buildings and structures that provide greater satisfaction, well-being and value to customers and users
- respect stakeholders and treat them fairly
- enhance and protect the natural environment
- minimise its impact on the consumption of energy (especially carbon-based energy) and natural resources.

The document states the need for a culture change within the industry and seeks to provide the catalyst for this change by promoting awareness and understanding of the issues and identifying priority areas for action.

Sustainable construction encompasses the complete life-cycle of a structure from initial concept through to demolition and site remediation. This involves all those that develop, plan, design, build, alter, or maintain the built environment and includes building materials manufacturers and suppliers as well as clients and end use occupiers (CBPP, 2000). In practice, delivering more sustainable construction involves all stakeholders within the industry, at each stage of the construction cycle, taking simple actions to deliver better projects (Figure 1.1).

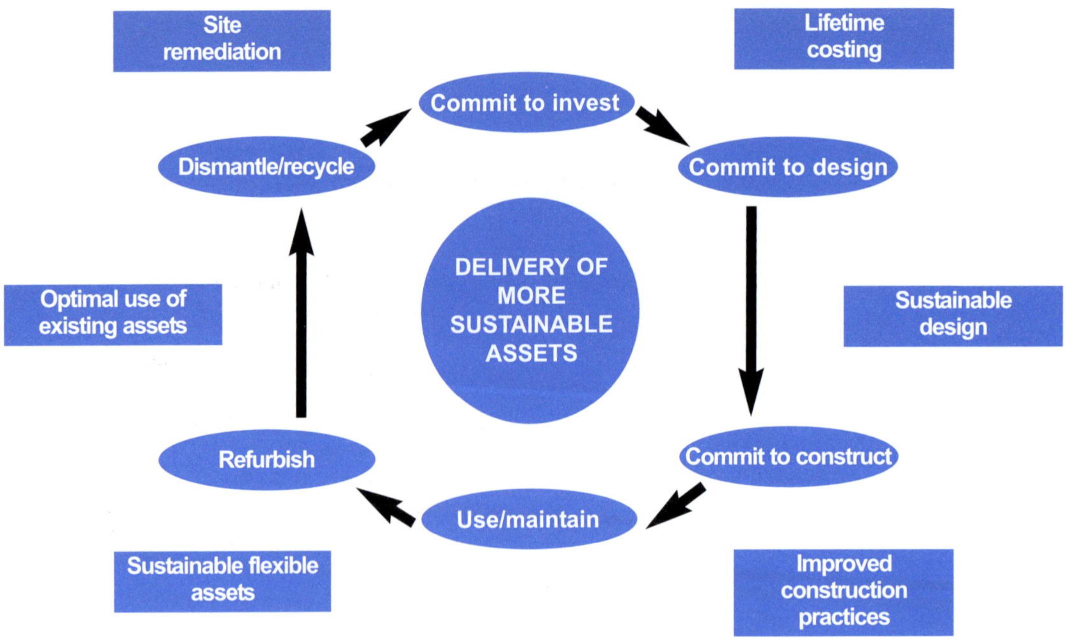

Fig 1.1 *Life-cycle of a sustainable construction project*

1.1.2 Key themes and issues

Within the four elements of sustainable development 10 key themes for sustainable construction have emerged from consultation within the construction industry. These themes, listed in Table 1.1 (which combines the first two elements), are particularly relevant for the development of company indicators which will demonstrate the value of measurement to assess and improve performance.

Table 1.1 *Key themes and issues for sustainable construction*

Principal issues

Effective protection of the environment and prudent use of natural resources

KEY THEME 1

Avoiding pollution
- Mitigation and management of pollution from site
- Transport planning

KEY THEME 2

Protecting and enhancing bio-diversity
- Habitat creation and environmental improvement
- Optimising use of brownfield sites
- Environmentally sensitive design and construction

KEY THEME 3

Improved energy efficiency
- Designing for whole life costs
- Use of local supplies and materials with low embodied energy

KEY THEME 4

Efficient use of resources
- Waste minimisation and management
- Re-use of existing built assets
- Lean design and construction
- Water conservation
- Use of recycled/sustaintably sourced products

Principal issues

Social progress which recognises the needs of everyone

KEY THEME 5

Respect for staff

- Provision of effective training and appraisals
- Equitable terms and conditions
- Provision of equal opportunities
- Health, safety and provision of conducive working environment
- Maintaining morale and employee satisfaction
- Participating in decision-making

KEY THEME 6

Working with local communities

- Minimising local nuisance and disruption
- Building effective channels of communication
- Contributing to the local economy
- Delivering structures and buildings that enhance the local environment

KEY THEME 7

Partnership working

- Building long-term relationships with clients
- Building long-term relationships with suppliers
- Corporate citizenship
- Delivering buildings and structures that provide greater satisfaction, wellbeing and value to clients and users
- Contributing to sustainable development globally

Principal issues

Maintenance of high and stable levels of economic growth and employment

KEY THEME 8

Increased productivity and profitability

- Improved productivity
- Consistent profit growth

KEY THEME 9

Improved project delivery

- Client satisfaction
- Minimising defects
- Shorter and more predictable completion time
- Lower cost projects with increased cost predictability

KEY THEME 10

Monitoring and reporting performance against targets

- Company reporting
- Benchmarking performance

Appendix 1 provides further details of how the industry can contribute to addressing each key theme of sustainable construction.

1.2 THE NEED FOR CHANGE

1.2.1 Industry performance

At present, the construction industry is widely perceived to be under-performing in each of the key sustainability theme groups: environmental, social and economic. In response, a culture of blame has developed in which each sector of the industry allocates responsibility for its current failings to others.

Rethinking Construction[1] (DETR, 1998) recognised that whilst, at its best, the industry can be innovative, adaptable and flexible, *'the construction industry too often fails to meet the needs of modern businesses that must be competitive in international markets, and rarely provides best value for clients and taxpayers'*. Adopting a sustainable approach would address this failure and lead to significant business benefits, including better understanding of the needs of clients, identification of opportunities for innovation, increased shareholder value, reduced costs, reduced risk, enhanced public relations and community liaison, and increased employee motivation.

The construction industry is a major polluter of the environment. In 1997, the industry was directly responsible for 588 water pollution incidents (four per cent of the national total) with fines for some individual incidents exceeding £10 000. The Environment Agency considers the current environmental performance of the industry to be unsatisfactory and in the medium-term future regulatory controls may therefore increase.

> **The construction industry has not done enough to reduce pollution and waste; to recycle material; it persists in taking too short a term view of the future and does not value the environment in all the business decisions it takes**
>
> **Ed Gallagher**
> **Chief Executive**
> **Environment Agency**

Effective management of environmental performance produces benefits by reducing the risk of non-compliance with regulatory requirements and hence minimises the financial penalties and negative publicity such breaches cause.

The construction industry is also a major consumer of natural resources. The annual net change from rural to urban land use is estimated to be 6500 ha, and 260 million tonnes of minerals are extracted per annum for aggregate and construction product manufacture (DETR 2000). The Aggregates and Landfill Taxes and changes to the Waste Management Licensing Regulations are increasingly making recycling of construction materials a commercial necessity. UK construction and demolition site waste currently accounts for 20-24 M tonnes each year and represents about six per cent of the total waste produced per annum in the UK. In total, about 30-40 per cent of construction and demolition waste is currently sent to landfill, presenting a considerable opportunity for further improvement in material recycling and cost savings.

Improving relationships with the public is equally important since public perception of the industry is currently generally negative. Construction is seen as damaging to the environment, dirty, disruptive and dangerous - and, given the high visibility of its rogue traders, it is also sometimes thought of as dishonest. Construction activities account for 12 per cent of all commercial and industrial noise complaints (DETR, 1998) and significant numbers of complaints regarding dust. The Urban White Paper (DETR, 1999c) and Urban Task Force are acting as drivers to improved industry performance. Individual companies and projects can also benefit from improved public relations and perceptions, for instance by acting to reduce any local opposition to new construction projects and thus minimising the delays and costs associated with securing planning permission.

[1]**Prepared by the Construction Task Force chaired by Sir John Egan**

Improving the public perception of the industry is also important in closing the skills gap which exists. The industry is projected to grow at 2.5 per cent per annum up to 2005. The Construction Industry Training Board (CITB, 1999) however estimates that there will be an annual shortfall of 6500 tradespeople in the building trades over this period. Training of staff will reduce this shortfall – but only to 2400 tradespeople per annum if training intake is not increased. The number of 16–24 year-olds within the workforce has declined to just 12 per cent. To meet projected demand the industry must attract and train more staff; and individual companies will need to differentiate their remuneration packages to recruit and retain the most able staff including more women.

It is impossible to identify and effectively target resources in order to address the unsustainable characteristics of current industry performance without quantitative data to indicate where improvements are necessary. Indicators of industry performance are therefore critical to both determine current performance and measure progress in addressing problems.

1.2.2 Business benefits of sustainable construction

The business case for more sustainable construction is not yet widely recognised across the industry. Nevertheless, enlightened industry leaders realise that, overall, the current performance of the construction industry is unsustainable and that change is inevitable. Government has signalled its intention to stimulate the required changes through fiscal and regulatory controls and by encouraging better practice through its role as a construction client.

> **Construction companies need to radically rethink their approach to sustainable construction. Companies that improve their environmental and social performance will receive increasing recognition from clients, investors, regulators and the wider community, leading to more repeat business, more efficient processes and higher margins. Indicators are an effective mechanism to help improve performance, by providing the practical tools to monitor performance, stimulate change, inform stakeholders, and benchmark performance against competitors.**
>
> **George Martin**
> **Business Affairs Director**
> **Forum for the Future**

Environment

Active management of sustainability performance can deliver significant improvements in business efficiency and profitability. In the environmental area, lean construction and pollution prevention can generate quantifiable cost savings, as can reduced energy consumption in business activities (during the construction process and in building design and operation). Examples of short-term business benefits that arise from improved environmental performance include:

- reduced landfill disposal and material purchasing costs by increasing recycling
- reduced transport costs through using local suppliers
- increased efficiency and better risk management for regulatory compliance through the introduction of environmental management systems
- positive publicity and an enhanced professional reputation arising from environmental improvement schemes, producing increased future tendering opportunities
- avoidance of pollution incidents or environmental degradation preventing associated fines, court costs and the related negative publicity.

Social

From a social perspective, there is strong evidence of economic benefits arising from better relationships with clients, the local community and other stakeholders, while evidence is also starting to emerge of productivity gains for staff involved in environmental and social performance improvement schemes, or as occupiers of sustainable buildings. Urban regeneration is another area in which construction makes an important contribution to sustainable development and can achieve positive community benefits.

Initiatives to improve relations with local communities affected by construction projects may also achieve significant business benefits. Involving local people in decision-making may bring additional local value to the project. Working with local communities can also reduce opposition to the project, enabling:

- planning permissions to be achieved more quickly and at reduced cost
- minimisation or prevention of direct action during the construction phase, hence reducing delays and site security costs
- enhancement of local reputation through positive publicity.

Initiatives to improve staff skills, job satisfaction and motivation also have positive business benefits in improved efficiency and productivity. The business benefits of a well-managed, skilled and motivated workforce, plus attractive working conditions, are significant: an increased capacity to innovate, an ability to recruit the most able individuals, and a capacity to retain able staff and so reduce recruitment and training costs and loss of institutional knowledge. Improved health and safety practices benefit business performance by improving efficiency and reducing accidents, saving on both management time and legal costs and fines.

Economic

The short-term financial benefits arising from improved partnering arrangements can also be significant. Rethinking Construction (DETR, 1998) highlighted the need for construction companies to develop more sustainable, long-term relationships with clients and suppliers and make greater use of partnering arrangements. Where partnering arrangements have been implemented cost savings of up to 50 per cent and time savings of 80 per cent have been reported for series of projects (DETR 1998, CIRIA 2000).

Many leading clients and investors are now demanding better environmental (and social) performance from their construction supply chain. For example, the Government Construction Clients Panel, which is responsible for 40 per cent of construction spend, has established a programme and timetable to deliver more sustainable construction in all Government procurement (GCCP, 2000). Improving project delivery is essential to build trust with clients, improve profitability and investment and encourage the development of mutually beneficial long-term relationships with clients.

Innovative, well-managed construction companies have recognised the need for change and are responding to the opportunities available to enhance or sustain their market position. They are also demonstrating their awareness of sustainability issues by publishing environmental and social reports to communicate this message to key stakeholders. Businesses that neglect their environmental and social responsibilities will lose market position and will find their activities increasingly constrained and frustrated by local objections and increasing requirements from clients and planning and regulatory authorities. The market conditions are now right to deliver more sustainable construction and will continue to improve as barriers to improvements are reduced or removed.

1.2.3 Steps to more sustainable construction

To achieve and access the benefits of more sustainable construction, action is needed at a number of levels including:

- initiatives across the industry and by industry sectors
- company actions to introduce new procedures and implement training
- delivery of more sustainable projects.

More sustainable construction also requires action throughout the life-cycle of construction projects, by all those involved in the process, to deliver improvements in the economic, social and environmental dimensions of sustainable development.

There are a number of practical actions that construction companies can take to move towards more sustainable construction. These include (DETR, 2000):

- make a commitment to address impacts – consider adopting a formal Environmental Management System
- aim to be best in class – benchmark your performance
- embrace producer responsibility – take account of whole life costs and impacts
- communicate with stakeholders – report on environmental and social performance
- work with others – improve the performance of your supply chain
- improve staff awareness through training to encourage innovation.

Individual companies are at different stages on the path towards sustainable construction and specific actions for any company will depend on their stage of development.

At a project level there are also a number of generic themes for action including:

- re-use existing built assets
- design for minimum waste
- design for re-use
- minimise energy in use
- preserve and enhance bio-diversity
- conserve water resources
- respect people and their local environment.

Case study 1 Page 65

...provides an illustration of sustainable office design – the new Arup Office Campus. This project has achieved significant energy and water savings compared with existing office space through setting environmental objectives and targets as part of the design process and adopting innovative approaches to meet them.

Using indicators to measure company and project sustainability performance and then identifying priority areas for improvement is one key aspect in improving sustainability performance.

1.3 INDICATORS OF SUSTAINABLE CONSTRUCTION

1.3.1 The role and benefits of indicators

Indicators are quantitative measures of performance with which progress can be assessed. Indicators provide measurements that can be used to:

- encourage take-up of better practice through monitoring
- stimulate change by setting performance targets
- facilitate dialogue with stakeholders by reporting progress
- demonstrate the outcome of initiatives being undertaken
- enable performance to be benchmarked
- enhance learning from experience by quantitatively analysing past performance.

As far as possible indicators should be:

- relevant (action oriented) and scientifically valid
- representative
- repeatable
- responsive to change
- reasonably simple to interpret.

From a practical perspective indicators also need to be based upon readily available data that is of adequate quality or that can be collated in a cost-effective manner and updated, ideally on an annual basis.

> **Sustainability indicators are a key component of the industry strategy for more sustainable construction. The indicators will eliminate a large amount of misunderstanding and mystique surrounding sustainability and enable consistent measuring and reporting of performance. They will also provide readily available information that will allow clients to become more discerning when specifying projects.**
>
> **Sir Martin Laing**
> **Chairman, John Laing Plc**
> **Chairman, Sustainable Construction Focus Group**

Each indicator provides a snapshot of progress in one of the important dimensions of sustainability. Considering the suite of indicators as a whole assesses overall progress. Increasing the number of indicators within each suite improves the comprehensiveness of the data-set but increases the demands of data collection and overall interpretation. It is therefore necessary to balance the need to be comprehensive against that of cost-effectiveness and ease of interpretation. Specific information on how indicators can be used by construction companies, along with the business benefits of doing so, will be provided Section 5.

1.3.2 Indicator framework

Delivering sustainable construction needs action at a variety of levels. In order to support this, indicators are needed to measure performance for each of these levels and at all stages of the construction life-cycle. Figure 1.2 shows the dimensions that need to be addressed by a suite of sustainable construction indicators.

Fig 1.2 *Sustainable construction indicator matrix*

It is not appropriate to develop a single comprehensive, coherent suite of indicators for the entire industry or companies within it. This is because the construction industry has multiple dimensions and diverse sectors and stakeholders, so many of the proposed measures would be irrelevant or inappropriately constructed for large parts of the industry. Separate suites of indicators, and related targets, are needed to monitor and promote change throughout the different sub-sectors of the industry. The Construction Industry Board was co-ordinating the preparation of an indicator framework to ensure that initiatives at all levels (project, company, sector and industry) and across all sectors (clients, architects, material producers etc) are consistent and simple to use.

Current initiatives include:

- industry indicators to be developed by the Construction Industry Board drawing upon work undertaken by CIRIA
- sectoral indicators under development by a range of organisations
- company indicators presented in sections 2–4 of this report developed by industry through the CIRIA project
- project indicators and economic key performance indicators developed by the Construction Best Practice Programme (CBPP, 2000)
- project indicators under development by the Movement for Innovation (M4I) including an index to measure project sustainability (BRE, 2000)
- indicators to measure client performance developed by the Government Construction Clients' Panel (GCCP, 2000).

Further information is provided in Appendix A.2 on how these initiatives are being co-ordinated to compile a comprehensive and complementary suite of indicators at different levels within the construction industry.

Development of company indicators for design and construction companies

CIRIA Research Project 609 has developed sustainability indicators for companies providing design and construction services. These indicators have been developed following extensive research and consultation with the industry and its stakeholders.

Initially, a preliminary suite of indicators was developed, drawing upon previous research. This preliminary work identified a large number of environmental and economic indicators but revealed that knowledge of social issues in the construction industry was much less well developed. A significant amount of new research and consultation was therefore undertaken to identify social issues and potential indicators. This activity was led by the New Economics Foundation, a respected sustainable development think tank.

The preliminary suite of indicators was tested by these project partner companies:

The testing process involved an initial review of the provisional indicators followed by selected data gathering to test the feasibility of the proposed measures. The pilot companies tested the relevance, appropriateness, ease of compilation and value of information generated for each of the proposed indicators.

Feedback from the piloting was used to improve the practicality of the indicators and the revised suites were presented in a consultation paper issued in May 2000 to more than 500 industry and stakeholder representatives. A consultation workshop in June, attended by representatives of companies from across the industry, planning and regulatory authorities, Government Departments and agencies and other stakeholder groups provided a further opportunity to comment on and improve the proposed measures. About 100 organisations responded to the consultation and their comments have been used to further refine the indicator suites.

The final stage in developing the company indicators has involved extensive review and comment by members of the project steering group and team. The final suite of indicators, and how companies can use these, are presented in this report.

See page 67 for more detail.

Case study 2

...describes the approach developed by CIRIA to develop company indicators for design and construction companies.

❍ Page 66

1.3.3 Company indicators

Building A Better Quality Of Life (DETR, 2000) identified that it was important for individual businesses to set, and monitor their progress towards, targets for more sustainable construction. This report presents the outcomes of an industry-led initiative to develop a suite of indicators to monitor the progress of companies specifically engaged in design and construction services.

Sections 2-4 of this report propose a suite of company indicators that can be used by businesses providing design or construction services in order to measure their:

- environmental performance
- social progress
- economic growth and project delivery.

Section 5 considers in detail the benefits of sustainability reporting and benchmarking of company performance and the role of targets. Section 5 also describes an action plan for companies wishing to develop and use indicators.

The industry umbrella bodies will promote the use of company indicators among their own members. As this project focuses on indicators for companies providing design and construction services, these indicators will be promoted by the Construction Industry Council (CIC) for design companies and by the Construction Confederation for contracting organisations. Each body will work with their members to develop and collect performance data relevant to member companies.

Clients also have a strong interest in the sustainability performance of their suppliers. In December 2000, the Confederation of Construction Clients launched a Client's Charter that commits leading clients to measure their performance (in-house and supply chain) against the nationally adopted KPIs. This commitment includes measurement against environmental and social indicators, once an agreed set is developed and adopted as an industry standard.

Clients also have a strong interest in the sustainability performance of their suppliers. In December 2000, the Confederation of Construction Clients launched a Client's Charter that commits leading clients to measure their performance (in-house and supply chain) against the nationally adopted KPIs. This commitment includes measurement against environmental and social indicators, once an agreed set is developed and adopted as an industry standard.

To be effective, company indicators must be relevant to business needs and action-oriented, so that they encourage implementation of better practice through internal monitoring of performance.

Case study 3

❍ Page 67

...describes the process by which Montgomery Watson tested the indicators presented in this study and the outcomes of their assessment.

In addition to providing information about the performance of the company the indicators also provide the basis for consistent company reporting on sustainability. This facilitates meaningful comparisons between companies and should stimulate improvements through competition. Specific company benefits of using indicators to measure progress are detailed in Section 5.2 and include:

- promoting implementation of better practice within the company, leading to more efficient and competitive company processes and the construction of more sustainable assets
- providing quantitative data to support company environmental and social reporting, to enable analysis of past performance, and enhance learning from these experiences
- facilitating development of targets against which to measure progress
- benchmarking performance against competitors.

For most types of businesses, including construction companies, there are two principal types of indicators (CIRIA, 1999):

- strategic indicators – measuring the processes in place to improve company performance
- operational indicators – measuring the sustainability of the activities undertaken by the company and its performance in delivering more sustainable projects.

Different projects require different operational indicators. For example, water conservation may be of prime importance for the design and operation of an office block, but is likely to be less relevant during the construction phase. Companies undertaking design or construction activities should therefore select only those operational indicators relevant to their business. That selection will depend upon the particular sectors of the construction industry within which it operates, the services provided by the company and the types of projects undertaken. Further information on strategic and operational indicators follows in sections 1.3.4 and 1.3.5.

1.3.4 Strategic indicators

Strategic indicators are generic and relevant to most construction companies, irrespective of the types of activities undertaken. Strategic indicators measure the systems and processes in place within companies to improve their performance, for example the presence of an environmental management system or an Investors in People or similar scheme.

The information needed for strategic indicators is usually available corporately and therefore they are less resource-intensive to compile. Strategic indicators are also highly appropriate for reporting progress as part of environmental reporting (see Section 5.2.2).

Some of the proposed strategic indicators measure performance normalised against total business activity such as the total number of projects undertaken or business turnover. It is recognised that for some large companies only a proportion of the projects undertaken will involve construction activities. For these companies information should be normalised against the activity of the relevant parts of the business.

1.3.5 Operational indicators

Operational indicators measure the performance of the company in delivering more sustainable construction projects. For example, operational indicators in the area of waste minimisation are:

- ratio of materials disposed of to materials supplied
- tonnes of wastes arising to landfill per £ turnover
- tonnes of hazardous wastes arising per £ turnover.

Overall company performance on each of the operational indicators can be measured by aggregating company performance on individual projects from project level indicators. The Movement for Innovation is currently developing a suite of project level indicators (see Section A.2.1). This report uses project indicators to develop company operational indicators, based upon aggregating project level information. It is therefore important to choose indicators of project performance that can be readily aggregated to give meaningful information at the company level.

It is not appropriate to develop a single comprehensive suite of operational indicators relevant to all undertaking design and construction activities, since companies deliver a diverse range of project types, covering a wide range of sustainability issues. Companies should therefore measure and report performance against operational indicators selected from the very wide range available as most relevant to the sectors of the industry in which they work.

1.3.6 Indicators and targets

Arguably, indicators by themselves are unlikely to bring about performance improvement. Targets are necessary to support indicators and enable assessment of progress towards more sustainable construction. They also encourage businesses and employees to allocate appropriate resources to achieve the desired rate of progress. Best practice with respect to targets is that they should be SMART:

- **S**pecific – to each indicator
- **M**easurable – therefore quantifiable
- **A**chievable – but challenging
- **R**ealistic – recognising the wide variety of progress throughout the industry
- **T**ime-bound – with a date for attainment.

Ideally, industry-wide targets should be developed against which all companies' or sectors' aspirations and progress can be assessed. In order for such targets to be developed there must be consensus from companies and organisations within the industry, and their stakeholders, on the value and compliance data that is proposed. This consensus can only be reached after baseline data is gathered to evaluate current industry performance.

At present there is minimal baseline data on company, sectoral or industry performance against social and environmental indicators from which to gauge a challenging but attainable rate of progress. There is also a wide disparity of views regarding the rate of progress towards more sustainable construction that is possible and desirable, and therefore to which companies and sectoral organisations are prepared to commit. This disparity of views exists within the industry, between different companies, and to an even greater extent among key stakeholders such as clients, social and environmental organisations and regulators.

The absence of consensus and baseline data makes development of industry-wide SMART targets a medium term objective that should follow from more widespread use and reporting of indicators to generate a baseline of data. However, it is important that individual companies (and where appropriate sectoral organisations) develop their own targets and measure and report against these to promote change within their own organisation. Section 5.2.3 provides guidance to companies on setting targets against the indicators. See also the examples of targets in use in Appendix A3.

2 Environmental indicators

2.1 KEY THEMES

Ten key themes have been identified as necessary for the delivery of sustainable construction and development of indicators (Section 1.1.2). Four of these themes relate to environmental performance:

- avoiding pollution
- protecting and enhancing bio-diversity
- improved energy efficiency
- efficient use of resources.

For each of these themes there is a range of principal issues (Table 1.1). Most of these issues are of direct relevance to, or can be significantly influenced by, the actions of companies providing design and/or construction services. These include:

- waste minimisation and management
- mitigation and management of pollution from site
- transport planning
- environmentally sensitive design and construction
- use of local supplies and materials with low embodied energy
- lean design and construction
- water conservation
- use of recycled and sustainably sourced products.

For each the above issues, company performance can be measured through the use of indicators. There are some sustainable construction environmental issues that are only directly within the remit of the client and over which, in practice, companies providing design and/or construction services have little influence. Examples include:

- optimising use of brownfield sites
- habitat creation and environmental improvement
- re-use of existing built assets
- designing for whole life costs.

For these issues, companies providing design or construction services may attempt to influence their clients but have limited influence on the ultimate outcome. In these circumstances it is not appropriate for the performance of companies providing design and construction services to be measured through indicators. The issues should rather be addressed by indicators of client company performance, and of the industry's performance overall.

Case study 4 Page 68

...describes how the Environment Agency has been achieving improved environmental performance on infrastructure projects it undertakes, such as construction of flood defences. This is being done through use of Environmental Action Plans containing specific indicators and targets of environmental performance.

2.2 STRATEGIC ENVIRONMENTAL INDICATORS

There are eight strategic environmental indicators listed in Table 2.1. Some of them also contain a social dimension, demonstrating the inter-relationship between sustainability issues. Most of these indicators are relevant to companies providing design and construction services and could usefully be included within their annual reporting.

Table 2.1 *Strategic environmental indicators*

THEME	INDICATOR
Effective protection of the environment and prudent use of natural resources	Percentage turnover of company operations with a formal or independently certified Environmental Management System to ISO 14001 standard or equivalent
	Percentage score calculated using the Business in the Environment (BiE) index of company environmental engagement
	Percentage of projects for which an environmental assessment has been undertaken and proposed environmental mitigation measures implemented
	Percentage of company sites operating under the Considerate Constructors or related scheme*
	Number of formal environmental or nuisance notices served due to construction activities*
	Percentage of services by value obtained from companies with a certified Environmental Management System to ISO14001 or equivalent
	Percentage of projects for which standards of environmental performance and social engagement have been formally agreed with the client
	Percentage of projects for which whole-life costs have been calculated and/or a life-cycle assessment has been undertaken and used in the design of the project or the method of construction and materials used

* Indicator may not be relevant to design companies that do not undertake site supervision activities.

2.2.1 Percentage turnover of company operations with a formal or independently certified EMS

The indicator measures the percentage (in terms of turnover) of the company operations that operate an environmental management system (EMS) to ISO 14001 or EMAS standards. It is relevant to all companies involved in construction and their suppliers. The indicator measures the extent to which a business has considered and addressed its environmental impacts within its procedures and practices. It does not directly measure environmental performance of the company but this would be achieved by other indicators in the suite.

Most large companies introduce an EMS in a phased manner throughout their operations. The indicator measures the extent to which this has been accomplished. Due to the cost, or other considerations, some companies choose to develop and implement an EMS without seeking its accreditation. Accreditation is preferable since it provides an independent check on the quality and effectiveness of the system in place. Where companies have introduced formal systems but not sought accreditation, however, the ensuing benefits need to be recognised. A formal EMS has been defined as one in which there is *documented appraisal of significant impacts and objectives and targets and for which internal environmental audits to assess compliance have been performed.* Where a company's systems are not fully accredited this should be made clear in any reporting.

It is recognised that, in time, all the activities of a company may operate under an accredited EMS. At present this is likely to be the case for a very small proportion of construction companies and it is unlikely that the indicator would become obsolete in the near future. However, at a future date it will be necessary to review it.

Some companies may have no part of their activities operating under a formal EMS. These companies should record zero percent but if choosing to report their performance could report progress to implementation such as by recording the proportion of operations (in terms of turnover) that have:

- documented appraisal of significant impacts
- documented objectives and targets.

2.2.2 Percentage score calculated using the BiE index of company environmental engagement

Index (BiE, 2000) measures the extent to which businesses take account of environmental issues in decision making. It is applicable to companies throughout the construction industry but is most relevant to larger businesses. It is compiled for FTSE 250 companies enabling comparisons both within and between sectors. The principal benefit of the index is that it allows simple benchmarking and comparisons of companies within the construction industry. The key limitation is that the score cannot be readily interpreted other than by comparison with those of other companies. There is also a small fee associated with determination of the company score.

In general indexes are not favoured since they lack transparency. This indicator does however provide an effective means of benchmarking both within the industry and between sectors.

2.2.3 Percentage of projects with environmental appraisal and environmental mitigation measures

There is a range of development projects which due to their size or locations are required (by the EC Directive on Environmental Impact Assessment Directive 97/11/EC) to undertake an environmental impact assessment (EIA). This is to determine whether the environmental harm is acceptable and how it can be mitigated or minimised. For the majority of construction projects there is no statutory requirement to undertake an EIA - but considerable business and environmental benefits flow from performing an environmental appraisal. Ideally this should be undertaken at an early stage of the project as part of the initial project conception or design. However, where this has not happened, there are benefits in undertaking an appraisal prior to the commencement of construction activities.

An appropriate scope for the appraisal will vary with the type of development and nature of the site and surrounding areas. It is beyond the remit of this study to provide detailed guidance but an appropriate approach is likely to be similar to that recommended by the European Commission on Scoping an EIA (EC, 1996) or, for buildings, possibly a BREEAM (Building Research Establishment Environmental Assessment Method) assessment. The objectives of the appraisal will be to identify significant environmental impacts and how these can be mitigated. It is likely to include consideration of impacts upon:

- the local community
- fauna and flora
- landscape
- material assets (such as heritage)
- soil, water, air and noise pollution.

As environmental controls become increasingly stringent, it makes good business sense to have the capacity to identify and address issues that may subsequently delay the project, or require costly redesign or amendments to the construction. An environmental appraisal will assist in securing necessary planning consents. It may also prevent delays and costs that would result from action to abate unacceptable noise or dust during construction or identifying protected species on site during the construction phase.

An important element of the scoping will include seeking the views of stakeholders, including the local community. The indicator therefore also contributes to delivering better working relations with this stakeholder, which is one of the key themes of sustainable construction. This specific issue is also addressed by social indicators (see Section 3).

The indicator measures the percentage of projects for which an environmental appraisal has been undertaken relative to the total number of construction projects performed. Performing an appraisal is irrelevant if the outcomes and recommendations are ignored. The indicator therefore also requires recommendations arising from the appraisal to be implemented. The absolute number of projects (as opposed to turnover) is used as the measure. There is likely to be a statutory requirement for an EIA for the largest projects. The indicator is designed to stimulate non-statutory appraisals whilst measuring the extent to which appraisals are undertaken overall.

This indicator is relevant to both design consultants and constructors, since the way in which both phases of the project are undertaken should reflect local circumstances. There is no need for separate or repeat appraisals to be undertaken, but where no appraisal has been performed during the design phase, best practice is that the constructor should undertake such an appraisal prior to commencement of works on site. This should prevent problems arising during the construction phase.

2.2.4 Percentage of company sites operating under the Considerate Constructors or related scheme

The Considerate Constructors Scheme (CCS, 2000) is a national, voluntary initiative operated by the Construction Confederation on behalf of the Construction Industry Board (CIB, 2000). Its aim is to improve the image of construction through better management and presentation of its sites by raising the standards of site design and management above statutory requirements. Constructors must commit to a Code of Considerate Practice that includes being:

- considerate
- environmentally aware
- clean
- a good neighbour
- respectful
- safe
- responsible
- accountable

Some of these requirements also embrace elements of social responsibility, particularly respect for the local community and staff. Experienced professionals who monitor the scheme visit sites and suggest and encourage improvements.

Individual projects registered through the scheme are eligible to join at any stage of the construction process from inception to completion. It is therefore relevant to both design consultants and contractors but mainly relates to the latter. The indicator measures the percentage of a company's sites that are registered with the Considerate Constructors or equivalent scheme. This enables sites that are registered with particular local authority schemes to be recognised within the indicator.

2.2.5 Number of environmental or nuisance notices served due to construction activities

This indicator measures the number of formal environmental or nuisance notices served by regulatory authorities on the company due to its activities or those of appointed subcontractors. It provides a simple measure of the effectiveness of the business in preventing environmental harm or nuisance through its operations and the negative impacts of these upon the business. The indicator is relevant to those involved in construction activities and may also be relevant to some design consultants, particularly those that undertake supervisory services on site.

The company should also measure the notices by type in order to assist with development of appropriate procedures to prevent reoccurrence. Notices could be categorised as those affecting or arising from:

- habitat or bio-diversity
- air quality emissions or dust
- noise
- waste disposal
- spillage to water
- river damage
- land contamination
- damage to heritage
- access rights
- other.

A related indicator is the *number of occasions the company was successfully prosecuted for breaches of environmental legislation and the value of fines incurred.* Relatively few companies are prosecuted and the indicator is therefore not particularly sensitive to changes in practices on an annual basis. However, for companies that choose to report their environmental performance any successful prosecutions against the company and the reasons and outcomes should be recorded.

2.2.6 Percentage of services by value obtained from companies with a formal or certified EMS

This indicator measures the extent to which the company integrates environmental performance within its supply chain management. It measures the value of services obtained from companies with a certified or formal EMS relative to those with no system. It is similar to the indicator for implementation of an EMS within the company itself. Supply chain pressure has been effective in other countries and industries to promote uptake of EMS and the expectation is that it would greatly enhance the performance of the UK construction industry.

As suggested in Section 2.2.1, accreditation of the EMS is preferable since it provides an independent check on the quality and effectiveness of the system in place. However, where suppliers have a formal uncertified system this should also be recognised (a formal EMS being defined in the same way as in Section 2.2.1). Where companies report the percentage of suppliers with an EMS they should also make clear the percentage that are certified. In reporting progress companies may also wish to indicate the progress of their suppliers to implement an EMS by reporting the proportion of its suppliers (in terms of turnover) that have:

- documented appraisal of significant impacts
- documented objectives and targets.

2.2.7 Percentage of projects for which standards of environmental performance have been formally agreed with the client

The previous indicator sought to promote improved environmental performance on projects through the supply chain. Another effective channel to stimulate improved environmental performance (and social engagement) on a project is to discuss opportunities with the client. Ultimately the extent to which companies providing design and construction services can deliver improved environmental (and social) performance on a project will depend upon the willingness of the client to act upon advice. All sub-contracting companies have a responsibility to raise awareness of potential opportunities with the client and to ensure that standards of performance have been agreed.

This indicator measures the percentage of projects for which standards of environmental performance (and social engagement) have been formally agreed with the client. This requires discussion and agreement on the approach to be used and the provision of documented evidence. For buildings, possible areas of discussion on environmental issues may include:

- the energy efficiency of the structure (beyond minimum building standards)
- choice of heating and cooling systems
- opportunities to use prefabricated materials or those with low embodied energy
- use of secondary materials
- landscaping or habitat creation opportunities
- environmental performance of subcontractors (such as the need to have an EMS).

Similar areas for discussion and agreement can be reached on social issues such as:

- the extent to which and how local communities are consulted
- engagement with end users
- employment conditions for site workers.

2.2.8 Percentage of projects for which whole-life costs have been calculated and/or a life-cycle assessment undertaken

Whole-life costing of projects takes account of the costs and resource demands in using and maintaining the asset in addition to those for its initial construction. Consideration of the whole-life costs of an asset as part of the initial design can mean considerable financial and resource savings whilst it is in use. For example, by designing a building to be energy efficient it is possible to make significant savings in energy bills. This indicator measures the proportion of projects for which companies consider whole-life costs as part of the initial design and construction process or as part of refurbishment.

Similarly, companies can also seek to use materials with a low environmental impact to reduce the energy, waste, pollution and other impacts associated with constructing the asset. The indicator also measures the proportion of projects for which life-cycle assessment considerations have been incorporated within the design and/or construction process.

2.3 OPERATIONAL ENVIRONMENTAL INDICATORS

2.3.1 Key issues

Operational environmental indicators measure the performance of the company in delivering construction projects in terms of their environmental impact both during the construction process and in the subsequent use of the built asset. Company performance is measured by aggregating the performance on individual projects.

Four key environmental themes have been identified for the construction industry (Section 1.2). Table 2.2 summarises the key issues specifically for companies providing design and construction services and the ways in which they can most effectively contribute to improving the environmental performance of the assets under construction.

For each theme (and type of activity), one of the issue is emboldened to highlight action that would make the greatest contribution to improving environmental performance. For design activities the priorities relate to measures that would improve environmental performance over the lifetime of the building. The priority issues for construction activities are the principal environmental issues of concern to regulators and the public.

The table is not an exhaustive list but focuses on priority issues for most companies. It does not address environmental issues that are largely beyond the scope of companies undertaking design or construction activities and are directly within the remit of the client or user, such as the proportion of designs or construction activities undertaken on greenfield sites.

Table 2.2 *Key environmental issues in design and construction activities*

	DESIGN ISSUES	CONSTRUCTION ISSUES
KEY THEME 1	**Avoiding pollution** ❍ **identification of potential pollution sources and design of mitigation measures** ❍ effective site supervision (where consultant is employed in a client resident engineer capacity)	❍ **preventing nuisance from noise and dust by good site management** ❍ minimising polluting emissions ❍ preventing pollution incidents and breaches of environmental requirements
KEY THEME 2	**Protecting and enhancing bio-diversity** ❍ **identification of opportunities to enhancing bio-diversity by habitat creation** ❍ design-in environmental improvements such as by landscaping or habitat creation	❍ **protecting sensitive ecosystems through good construction practices and supervision**
KEY THEME 3	**Efficient use of resources** ❍ propose lean designs avoiding over-engineering ❍ **design for whole life-costs and minimum waste in construction, use and afterlife** ❍ specify use of recycled materials, or those from sustainable supplies or with low embodied energies ❍ design to conserve water in use including, where appropriate, rainfall-capture and grey-water systems	❍ **lean construction avoiding waste** ❍ use recycled materials, or those from sustainable supplies or with low embodied energies ❍ conserve water during construction
KEY THEME 4	**Improved energy efficiency** ❍ **energy efficient building design** ❍ reduced energy consumption in business activities ❍ incorporation of renewable energy sources or combined heat and power schemes into development design	❍ energy efficient construction processes ❍ **reduced energy consumption in business activities**

2.3.2 Role and limitations

The general role and limitations of operational indicators were discussed in Section 1.3.5. Company performance measured by operational environmental indicators will, in general, be influenced by factors beyond the direct control of the company including local project circumstances, and the attitude of the client. Choice of materials, for instance, is in part a client decision and they may specify, say, primary stone or hard woods that cannot be sourced from sustainable supplies. The operational environmental indicators provide companies with information about their overall performance in achieving more sustainable construction and act as a tool to aid discussions with the client.

As the company environmental performance on projects is not entirely within its control some companies may not wish to publicly release information about its performance measured against relevant operational environmental indicators. Companies are nevertheless encouraged to monitor performance against the indicators and evaluate progress through benchmarking against other companies. This is discussed further in Section 5.2.4.

Sections 2.4 and 2.5 list operational environmental indicators for design consultancy services and construction activities. Not all of the indicators listed in the following sections are relevant to all companies. Companies are therefore encouraged to select and use operational environmental indicators that are most relevant to the projects on which they work. The suites of indicators proposed are not comprehensive and further indicators may be useful to companies undertaking specialist services especially for infrastructure projects and in relation to resource efficiency.

Development of comprehensive suites of measures needs to be undertaken in parallel with strategies to improve the sustainability of each of the sub-sectors within the industry. An M4I project to develop Environmental Performance Indicators (EPIs) is in progress (see Section A.2.1).

Prescriptive approaches to quantify project performance are not included. This reflects the diversity of indicators and companies, and will hopefully encourage companies to adapt existing reporting procedures and develop the most cost-effective ways to measure and improve performance. The absence of a formal measurement and reporting procedures will reduce the consistency of any reported data. The initial priority, however, is to encourage companies to measure performance against the Operational Indicators and seek to improve their own performance. Once this is established more consistent measurement can be encouraged to enable effective benchmarking and reporting. This is discussed further in Section 5.2.

2.4 OPERATIONAL ENVIRONMENTAL INDICATORS FOR DESIGN CONSULTANTS

Operational environmental indicators against which companies providing design and consultancy activities can monitor their environmental performance on projects are listed in Table 2.3. Operational indicators are suggested for all key issues. For some indicators companies may wish to produce separate performance measures on different types of project, for example:

- infrastructure projects
- housing
- retail
- commercial.

Table 2.3 *Operational environmental indicators for companies providing design and consultancy services*

	ISSUE	INDICATORS
KEY THEME 1	**Avoiding pollution**	
	identification of potential pollution sources and design of mitigation measures	percentage of projects per £ turnover including: ❍ traffic management schemes during construction and use ❍ dust control ❍ noise controls ❍ measures to prevent ground contamination ❍ measures to prevent water contamination
	design for minimum waste in construction, use and afterlife	percentage of projects with a waste minimisation strategy
	effective site supervision (where consultant is employed in a client resident engineer capacity)	number of environmental or nuisance notices served due to contruction activities per £ turnover
KEY THEME 2	**Protecting and enhancing biodiversity**	
	design-in environmental improvements such as by landscaping	cost of environmental improvements designed as a proportion of total project costs
	identification of opportunities to enhancing biodiversity by habitat creation	percentage of project sites for which an assessment of existing biodiversity has been undertaken prior to design and appropriate mitigation measures implemented area of habitat designed and created
KEY THEME 3	**Efficient use of resources**	
	propose lean designs avoiding over-engineering	**no generic indicators – sector specific indicators are more appropriate**
	design for whole life costs	percentage of projects for which life-time costs have been derived and were a material consideration in the design
	specify use of recycled materials, or those from sustainable supplies or with low embodied energies	percentage of secondary and recycled aggregates used in construction percentage of timber used in construction from well managed sustainable sources
	design to conserve water in use including, where appropriate, rainfall-capture and grey-water systems	predicted treated water consumption for new buildings in litres per person per day (lpppd) averaged over all projects of a similar type design capacity of grey-water and rainfall collection systems

Continued on next page

KEY THEME 4

ISSUE	INDICATORS
Improved energy efficiency	
energy efficient building design	energy efficiency of the building design for domestic buildings – the aveage SAP rating for all dwellings designed energy efficiency of the building designs for non domestic buildings – average percentage beyond building regulations (either for insulation levels of the building fabric or an overall energy target) energy efficient construction processes percentage of projects (by turnover) incorporating renewable energy systems or combined heat and power
reduced energy consumption in business activities	CO_2 emitted per £ turnover (DETR, 1999b)
incorporation of renewable energy sources or combined heat and power schemes into development design	percentage of energy demand by structure provided from: ❍ renewable sources ❍ combined heat and power
Generic	percentage of projects (by turnover) for which the M4I index of project sustainability has been calculated. The average score of projects measured percentage of relevant projects (by turnover) for which a BREEAM score (or similar system) has been calculated percentage of projects achieving an "excellent" rating using BREEAM (or similar system) for new construction projects percentage of projects achieving an "excellent" rating using BREEAM (or similar system) for refurbishment projects

2.5 OPERATIONAL ENVIRONMENTAL INDICATORS FOR CONSTRUCTION ACTIVITIES

A separate suite of operational environmental indicators against which companies providing construction services can monitor their environmental performance on projects are listed in Table 2.4. As with those for design and consultancy services, companies may wish to produce separate performance measures on different types of project:

- infrastructure projects
- housing
- retail
- commercial.

Table 2.4 *Operational environmental indicators for companies providing construction services*

ISSUE	INDICATORS
Avoiding pollution	
minimising polluting emissions	SO_2, PM_{10} and NO_x released (in tonnes) per £ turnover arising from construction activities
	average distance travelled per tonne of materials from suppliers to site (km)
preventing nuisance from noise and dust by good site management	number of complaints received per project site for noise and dust
	proportion of construction costs on nuisance mitigation
construction waste minimisation and elimination	tonnes of wastes arising to landfill per £ turnover
	tonnes of hazardous wastes arising per £ turnover
preventing pollution incidents and breaches of environmental requirements	see strategic indicators
Protecting and enhancing bio-diversity	
protecting sensitive ecosystems through good construction practices and supervision	percentage of project sites for which appropriate mitigation measures have been implemented to protect sensitive ecosystems
	number of non-conformance certificates issued by £ turnover
Efficient use of resources	
lean construction avoiding waste	tonnes of unused construction materials disposed to landfill
	percentage of materials intentionally over-ordered by value
	percentage of recycled and secondary aggregate used in construction
use recycled materials, or those from sustainable supplies or with low embodied energies	percentage of timber used in construction from well managed, sustainable sources
conserve water during construction	water consumption in (m^3) per £ turnover arising from construction site activities

Continued on next page

ISSUE		INDICATORS
Improved energy efficiency		
	energy efficient construction processes	CO_2 released (in tonnes) per £ turnover arising from construction activities (DETR, 1999b)
	reduced energy consumption in business activities	CO_2 released (in tonnes) per £ turnover (DETR, 1999b), arising from business-related non-site activities
Generic		percentage of projects (by turnover) for which the M4I index of project sustainability has been calculated. The average score of projects measured

3 Social indicators

3.1 KEY THEMES

Social indicators examine the aspect of a company's performance that is to do with relationships with people and communities, and with building relationships with stakeholders, creating opportunities at a local level and on the corporate scale.

There is growing acceptance that social responsibility is a core business requirement for construction companies - and furthermore, that building trusting relationships with stakeholders and dcvcloping mutual understanding both generate new opportunities and help to manage risk. Section 1.2.1 identified three key social themes to sustainable construction:

- respect for staff
- working with local communities
- partnership working.

These themes encompass the critical groups with whom companies need to develop strong relationships.

3.2 STRATEGIC SOCIAL INDICATORS

The proposed strategic social indicators measure the effectiveness of the company's relationships with people, primarily its staff, the local communities upon whom its projects impact and business partners. They focus upon the impact of the company's policies and procedures and the process by which it interacts with these groups.

Most of the strategic social indicators are based upon measuring the frequency with which companies engage in good practice for individual projects or project sites. At present, few companies compile much if any of the required information demonstrating the progress to be made in consideration of social issues as against environmental or economic ones. To manage the effort required in data collection and compilation it is therefore proposed that large companies initially select representative samples of their projects or sites, or undertake annual surveys, to compile the required data.

There are eight strategic social indicators for construction companies. All strategic social indicators are relevant to all companies undertaking design and construction services. Some of the strategic environmental indicators that seek to reduce the nuisance caused by construction activities are also good measures of social performance.

Table 3.1 *Strategic social indicators*

INDICATOR

KEY THEME 5

Respect for staff

- percentage of staff covered under the terms of an Investors in People or similar scheme
- percentage of annual staff turnover for permanent staff
- percentage of services by value obtained from companies operating and Investors in People or similar scheme
- reportable fatal and non-fatal accidents per 100 000 hours worked

KEY THEME 6

Working with local communities

- percentage of projects that include (and implement) a plan for stakeholder dialogue
- percentage of appropriate projects that include (and implement) a plan to consult with the end user

KEY THEME 7

Partnership working

- percentage of turnover generated by projects undertaken under alliances or other forms of partnership working
- average client satisfaction using the KPI 1–10 scale

3.2.1 Percentage of staff covered under the terms of an Investors in People or similar scheme

Investors in People (IIP) is the national standard which sets a level of good practice for training and development of people to achieve business goals (IIP, 2000). IIP was developed by the National Training Task Force in partnership with leading national business, personnel, professional and employee organisations. The IIP Standard is based on four key principles:

- **commitment** – to invest in people to achieve business goals
- **planning** – how skills, individuals and teams are to be developed to achieve these goals
- **taking action** – to develop and use necessary skills in a well defined and continuing programme directly tied to business objectives
- **evaluating** – outcomes of training and development for individuals' progress towards goals, the value achieved and future needs.

IIP provides a standard against which to demonstrate respect for staff and is in use throughout the construction industry (CC, 2000). Most large companies introduce IIP in a phased manner throughout their operations. The indicator measures the extent to which this has been achieved. Similar schemes to IIP exist (although are less well established); the indicator is therefore not limited to a single scheme. IIP is equally relevant to large or smaller businesses.

3.2.2 Annual staff turnover for permanent staff

Staff turnover provides a measure of staff satisfaction with the employer. High staff turnover also creates a considerable cost to the business and therefore needs to be restricted to acceptable levels. The indicator measures the percentage of permanent full and part-time staff voluntarily leaving the company on an annual basis.

3.2.3 Percentage of services by value obtained from companies operating Investors in People or a similar scheme

This indicator measures the extent to which the company integrates social responsibility within its supply chain management. It measures the proportion of services obtained (by value) from companies that operate an IIP type system. Supply chain pressures have been effective in many countries and in the UK retail sector in promoting green procurement. This indicator is intended to measure and stimulate improvement in the way suppliers treat and train their staff.

3.2.4 Reportable fatal and non-fatal accidents per 100 000 hours worked

On average one construction worker dies at work as the result of an accident each day and it is estimated £180 million a year could be saved in work related illness costs in the construction industry. Better health and safety also makes sound business sense since it reduces:

- legal costs of prosecutions and civil claims
- insurance premiums
- sickness and injury leave.

The indicator measures the number of reportable fatal and non-fatal accidents per 100 000 hours worked on site. This is the standard measure employed by the HSE and includes separate values for fatalities and injuries leading to more than three days' absence from work.

3.2.5 Percentage of projects that include (and implement) a plan for stakeholder dialogue

At the core of working with local communities is the need for dialogue with the different stakeholder groups including:

- neighbours
- local community groups
- wider community.

Each of these groups may have different concerns and channels through which they can most effectively be reached (NEF, 1998). To engage with local communities a plan for stakeholder dialogue should be developed and implemented from the conception of the project. Where no plan has been prepared prior to the design or construction phases of the project it is best practice for those involved at these stages to seek to engage with the local community.

The plan should identify local stakeholders and how to engage with these. At its simplest this could involve sending letters to local residents providing information about the project, where to find more information and channels through which concerns or complaints can be made. A more advanced plan could potentially involve local meetings or participatory workshops. Higher levels of involvement would include inviting local representatives to project meetings or even involving local people on the team. The indicator is not prescriptive as to what forms of participation should be undertaken but seeks to measure and encourage local community involvement in projects.

3.2.6 Percentage of appropriate projects that include (and implement) a plan to consult with the end-user

This is analogous to the previous indicator but seeks to encourage and measure the extent to which end users are consulted. For some types of projects, such as house-building, end-user consultation with potential buyers is an important commercial consideration. For others such as infrastructure projects office workers, shoppers or members of the public as travellers are consulted much less widely about potential developments and more rarely input ideas into their conception and design. Ideally end-user consultation should be undertaken at an early stage of the design, but where this has not happened it can still be relevant prior to construction – particularly concerning how to minimise disruption.

3.2.7 Percentage of turnover generated by projects undertaken under alliances or other forms of partnership working

This indicator is less developed than the other strategic indicators suggested within the report. Rethinking Construction identified building alliances and long-term relationships between clients, designers and contractors as essential to improved performance. The indicator measures the proportion of turnover generated by projects undertaken under partnership or other forms of alliance. It thus embraces all work that is undertaken under framework or multi-project agreements rather than on a single tender basis for an individual item of work. While it is recognised that many types of contractual agreements exist, this broad definition should enable meaningful comparisons to be produced.

3.2.8 Average client satisfaction with the service of the consultants and main contractor for projects completed in 2000

This indicator adopts the KPI approach to assessment of client satisfaction, using a 1 to 10 scale where:

- 10 = totally satisfied
- 5 = neither satisfied nor dissatisfied
- 1 = totally dissatisfied.

Companies undertaking design and construction services could produce values for both services and an overall aggregate value.

3.3 OPERATIONAL SOCIAL INDICATORS

3.3.1 Key issues

Three social themes have been identified as key to the delivery of sustainable construction (Section 1.1.2):

- respect for staff
- working with local communities
- partnership working.

Within the scope of partnership working there are a number of important issues including:

- building long-term relationships with clients
- building long-term relationships with suppliers
- delivering buildings and structures that provide greater satisfaction, wellbeing and value to clients and users
- corporate citizenship.

Table 3.2 lists principal issues for monitoring the performance of companies providing design and construction activities.

Table 3.2 *Operational social issues*

Principal issues

KEY THEME 5

Respect for staff

- provision of effective training and appraisals
- equitable terms and conditions
- providing equal opportunities
- health, safety and provision of a conducive working environment
- morale and employee satisfaction
- participation in decision making

KEY THEME 6

Working with local communities

- minimisation of local nuisance and disruption
- building effective channels of communication
- contributing to the local economy
- corporate citizenship

KEY THEME 7

Partnership working

- building long-term relationships with clients
- building long-term relationships with suppliers

Section 1.1.2 identified a number of key issues encompassed within the partnership working theme. Of these building long-term relationships with clients and suppliers are directly within the scope of any construction company. Delivering buildings and structures that provide greater satisfaction, wellbeing and value to clients and users are primarily issues for clients. Contributing to sustainable development globally is primarily an issue for Governments, working with representative construction sector organisations, rather than specific companies.

3.3.2 Operational social indicators

Table 3.3 lists possible operational social indicators of relevance to construction companies providing design or construction services. As with the operational environmental indicators (Section 2.3) not all may be relevant to all companies. Companies are therefore encouraged to select and use operational social indicators that are most relevant to the projects on which they work. The suites of indicators proposed is not comprehensive and further indicators may be useful. Prescriptive approaches to quantify performance are not included. This reflects the diversity of indicators and companies, and will hopefully encourage companies to adapt existing reporting procedures and develop the most cost-effective ways to measure and improve performance.

Development of social indicators for the construction industry is at an early stage and the proposed measures are not therefore comprehensive. However, they do provide a good starting point from which companies can begin to improve their performance.

Table 3.3 *Operational social indicators*

Respect for staff	INDICATOR
Provision of effective training and appraisals	percentage of staff receiving formal annual appraisals percentage of staff expressing satisfaction with the appraisal system percentage of staff undertaking structured training programmes including social and environmental aspects of construction percentage of staff time used for giving or receiving formal training including social and environmental aspects of construction
Equitable terms and conditions	percentage of staff with a pension to which the company contributes percentage of part-time workers percentage of staff working with flexible hours percentage of staff offered flexible benefits percentage of staff working more than 48 hours per week number of employees opted out of the Working Time Directive percentage of employees declining offers of employment
Providing equal opportunities	proportion of women employed proportion of staff from ethnic minorities proportion of staff registered disabled proportion of women in senior management positions
Health, safety and provision of a condducive working environment	see Strategic indicators for health and safety number of site employees provided with showers and rest areas in place of work
Morale and employee satisfaction	percentage of staff involved in ongoing surveys of job satisfaction percentage of staff expressing satisfaction with the way the company treats them
Participation in decision making	percentage of staff represented through consultative committees percentage of eligible staff taking-up share-save or similar schemes percentage of company owned by employees

Continued on next page

Table 3.3 *Operational Social Indicators (Continued)*

Working with local communities	INDICATOR
Minimisation of local nuisance and disruption	see Strategic Indicators and operational environmental indicators number of complaints received about inappropriate behaviour from employees working on site
Building effective channels of communication	see Strategic Indicators
Partnership working	
Building long term relationships with clients	number of clients with whom long-term strategic alliances have been formally agreed
Building long term relationships with suppliers	number of suppliers with whom long-term strategic alliances have been formally agreed
Corporate citizenship	value of charitable donation in money or time as a proportion of profits[1]

[1] This is intended to measure the extent to which companies freely donate staff time or money to causes from which there is no direct business benefit.

4 Economic indicators

4.1 KEY THEMES

Rethinking Construction (DETR 1998) described an under-achieving industry, poorly perceived by many of its clients and investors, that was trapped in a cycle of unsatisfactory project delivery, under-investment and, in some sectors, low profitability. The Rethinking Construction Task Force proposed ambitious targets and challenged the industry to measure its performance against these. To provide a consistent basis against which companies and the industry could measure its performance the Key Performance Indicator (KPI) Working Group was established. It reported in January 2000 (DETR, 2000a).

The KPIs have been developed by the Construction Best Practice Programme (CBPP, 2000) to provide a consistent basis against which to measure business profitability and productivity. This has been achieved by development of a series of charts and other tools (CBPP, 2000a) to enable company performance to be benchmarked against that of others in the same sector. In addition to project delivery the KPIs also provide a consistent basis against which to measure business profitability and productivity.

The proposals for the industry within Rethinking Construction address the economic aspects of more sustainable construction and also encompass several issues addressed by the respect for staff and partnership working themes. This study has utilised appropriate KPIs as economic measures of more sustainable construction under two key themes:

- improved project delivery
- increased productivity and profitability

4.2 STRATEGIC ECONOMIC INDICATORS

The KPIs provide appropriate measures of company economic performance. In order that these can be presented in a manner consistent with the environmental and social indicators the KPIs have been divided into strategic and operational indicators in which the former measure overall company performance and the latter aggregated performance on projects. Other KPIs have been incorporated within the social indicators. Table 4.1 lists the strategic economic indicators.

Table 4.1 *Strategic economic indicators*

Improved project delivery	the average normalised time to construct a project less the normalised time to construct a project in 1999, expressed as a percentage of the latter
Increased productivity and profitability	profit (before tax and interest) as a percentage of sales profit (before tax and interest) per employee

These indicators use the KPI methodologies (CBPP, 2000). Details regarding calculation methods may be found within these publications.

4.3 OPERATIONAL ECONOMIC INDICATORS

The KPIs concerned with improved project delivery and increased profitability and productivity are listed in Table 4.2 (CBPP 2000). Targets for improving performance were established for these indicators as part of the process of developing the KPIs.

Table 4.2 *Operational economic indicators*

Theme	INDICATOR
Improved project delivery	average actual duration at Commit to Construct less the estimated duration at Commit to Invest, expressed as a percentage of the latter average actual duration at Available for Use less the estimated duration at Commit to Construct, expressed as a percentage of the latter average actual cost at Available for Use less the estimated cost at Commit to Construct, expressed as a percentage of the latter average actual cost at Available for Use less the estimated cost at Commit to Invest, expressed as a percentage of the latter average condition of the facility with respect to defects at the time of handover for projects completed, using a 1 to 10 scale average client satisfaction with the finished product or facility, for projects completed, using a 1 to 10 scale
Increased profitability and productivity	average normalised construction cost of a project less the normalised cost of a similar project in 1999, expressed as a percentage of the latter company value added per employee (£) (Value added is turnover less all costs subcontracted to, or supplied by, other parties) profit before tax and interest as a percentage of sales, reported in 2000

5 Using company indicators

5.1 INTRODUCTION

Sections 2 to 4 identified a range of environmental, social and economic indicators of sustainable construction for design and construction companies. This section outlines how these companies can specifically use indicators to improve the sustainability of their businesses and the specific benefits of using indicators. Section 5.2 highlights the specific company benefits of using indicators to measure progress:

- promoting implementation of better practice within the company leading to more efficient and competitive company processes and construction of more sustainable assets (Section 5.2.1)
- providing quantitative data to support company environmental and social reporting and to enable analysis of past performance and enhance learning from these experiences (Section 5.2.2)
- facilitating development of targets against which to measure progress (Section 5.2.3)
- benchmarking of performance against competitors (Section 5.2.4).

Section 5.3 identifies the key next steps for companies wishing to make use of indicators to measure and enhance their performance, including data compilation, performance improvement measures and reporting. Section 5.4 provides a checklist of the principal actions companies need to take to deliver improved performance.

5.2 REAPING THE REWARDS

5.2.1 Promoting implementation of better practice

Strategic indicators

Monitoring of business performance is a key management responsibility. The proposed strategic indicators enable management to measure progress with elements of company environmental and social performance in addition to standard financial indicators. Specifically, the strategic indicators encourage adoption of better practice by monitoring, and promoting uptake of, business systems that themselves facilitate improved performance. For example, the strategic indicators enable the company to examine the progress being made in the introduction of environmental management systems or Investors in People and similar schemes for its own and supplier operations.

Page 70

Case study 5

...describes the experiences of the WSP Group in initiating a group-wide sustainability initiative to integrate the principles of sustainability within all their activities. This has included setting indicators to measure the effectiveness of the changes that have been implemented in management processes.

Monitoring of company progress acts as a stimulus to those responsible for making progress in each of the areas, particularly if targets have also been set or performance is benchmarked against other companies (see the following sections).

Operational indicators

The operational indicators measure the extent to which companies are effective in implementing more sustainable construction practices on the projects they undertake. The operational indicators therefore provide a direct measure of company progress towards more sustainable construction, complementing the strategic indicators.

The use of operational indicators requires information to be generated at a project level. This project level information is then collated to produce aggregate company data. The generation of data at the project level focuses the attention of project managers (or whoever is charged with providing the required information) on the particular sustainable construction issue and will in turn influence their management of the project or activity.

As far as possible, those staff with responsibility for providing the information should also be those whose decision making will ultimately lead to improved implementation of better practice. Where possible, the project level information used to compile the operational indicators should therefore be produced during the project, not retrospectively. This enables the project manager to proactively seek to improve performance on the project whilst it is ongoing.

> **It is vital that the industry improves the environmental and social performance of its projects. However, sustainability can be confusing and put people off, creating a barrier to performance improvement. Project teams need a simple set of indicators to highlight the key performance issues they should address and to enable them to measure their own performance, benchmarked against best practice or industry standards.**
>
> **Rab Bennetts**
> **Bennetts Associates Architects**
> **Chairman, Movement for Innovation**
> **Sustainability working party**

To facilitate efficient data collection, procedures to compile and collate the necessary information should be integrated with established quality or environmental management systems. Once the data from a cross section of projects has been collated the analysis of the information provides a powerful tool with which the company can assess the performance of different sections of the business or individual project managers. Where companies use operational indicators as measures to assess individual project manager performance the focus to improve project delivery is especially clear.

The collated information from a range of projects can also be used to identify clients that are particularly receptive to sustainable construction initiatives in order to add value to future tenders.

Case study 6 ❍ Page 71

...describes an innovative approach being used by ICI to measure and cost-effectively improve their environmental performance using an integrated environmental index linked to the company environmental objectives.

5.2.2 Use of indicators in business reporting

In addition to monitoring and stimulating improved performance, indicators also provide good means for reporting of company performance. To be effective, and meaningful, reporting must be part of an effective communication strategy and not merely a public relations exercise. Reporting of company performance against indicators enhances the credibility and effectiveness of company environmental and social reports that in turn:

- demonstrate the outcomes of the initiatives being undertaken
- demonstrate to regulators and government that the company is complying with requirements and is at the forefront of initiatives to improve performance helping to build effective working relationships
- demonstrate that the company is responding to clients' concerns and needs
- encourage employees to apply good practice and add urgency to the need for change
- facilitate dialogue with stakeholder groups
- build trust of the local community and neighbours, making subsequent expansion or changes to operating practices easier to achieve
- inform the wider community through the company's own words and actions and not those of journalists and pressure groups that may misrepresent the company
- demonstrate to clients, lenders, investors, insurers etc that the business is well-managed and is seeking to reduce risks and minimise future liabilities.

Several major construction companies have already prepared environmental reports (CIE, 2000) and these contain a range of quantitative information to assess impacts and performance.

Case study 7
...provides details of the John Laing Sustainability Report.

Page 72

The DETR (2000b) has provided useful guidance on environmental reporting with links to other sources of information. Much of this is also relevant to social reporting. The indicators proposed in this document provide valuable quantitative data to demonstrate company progress on an annual basis.

Whilst companies are obviously free to report against whatever indicators they consider appropriate, by reporting against each strategic indicator companies will be able to benchmark their performance against competitors and assist in improving standards throughout the industry. Reporting against each strategic indicator will also prevent accusations of companies selectively cherry-picking the indicators against which they choose to report, reducing the credibility and effectiveness of the reporting process.

Operational indicators

The operational indicators also provide valuable information about project performance. The indicator performance on an individual project is however affected by local circumstances and by the attitude of clients. Both have a strong influence upon the approach adopted during the project and the extent to which it is possible to incorporate best practice for more sustainable construction. Accordingly, an individual company may be able only partially to influence its sustainability performance on a particular project; and this will feed through, to some extent, in its performance measured by individual operational indicators in a particular year. Some companies may therefore choose not to publicly report performance as measured against operational indicators, but rather to use these for internal monitoring or benchmarking performance against other companies.

Some operational indicators can effectively be included in company reports and should be assessed to determine the value of so doing. In practice, in most instances, only the larger consultancy and contracting companies are likely to have the time and resources to prepare environmental reports. These larger businesses undertake the greater share of projects in terms both of value and of carrying the greatest environmental and social implications.

SMEs

SMEs can make a significant contribution to more sustainable construction. The resources, market and priorities of these businesses are however different from those of larger consultancy and contracting companies, and they are therefore unlikely to produce environmental reports. Indicators are nevertheless relevant to SMEs because they also will be able to improve their sustainability performance, through improved information and training, tighter regulation and via the subcontracting system. The uptake of these activities can be monitored through sectoral indicators and company indicators linked to procurement of services from subcontracting companies.

5.2.3 Use and development of targets

Setting targets for each indicator provides a further incentive for improvement. Section 1.3.6 described best practice for setting targets. In setting targets, companies should seek to achieve challenging but achievable rates of change that reflect the resources that are to be allocated to facilitate progress.

Most companies will, at least initially, choose to develop their own targets for each of the indicators reflecting individual priorities, circumstances and available resources. A range of national targets has been established that companies may find useful for developing their own targets. There is also a range of targets published by other organisations including:

- international organisations
- non-governmental organisations
- industry bodies
- professional organisations
- other companies in environmental reports.

Appendix A.3 lists some relevant criteria against which companies could develop their own targets for environmental and social issues. The Construction Task Force (DETR, 1998) has proposed targets for each of the economic indicators. These are listed in Table 4.2.

In the medium term it should be possible to develop industry-wide targets against which the progress of different companies can be measured. However, to do this there must first be a degree of consensus within the industry and with its stakeholders regarding what can be achieved and a baseline of data from which to assess recent performance. (Section 1.3.6).

5.2.4 Benchmarking performance

Another benefit of sustainability indicators is that they allow companies to benchmark performance against competitors through benchmarking clubs (CBPP, 2000a). The operational indicators identified in this report provide suitable measures for use by industry benchmarking clubs. Through the clubs, standardised procedures could be developed to measure performance against the indicators. Indicators and procedures are likely to be developed and refined with experience. It is hoped that the indicators presented in this report will therefore be adopted and adapted by benchmarking clubs so that companies assess their performance against the important sustainable construction issues.

Once operational indicators are reported, or adopted by benchmarking clubs, the current best-in-class will become the de-facto target against which other companies will aspire. In time, as use of the indicator increases, individual companies, through benchmarking clubs, may consider developing their own targets.

5.3 USING INDICATORS: NEXT STEPS

Delivering more sustainable construction requires action by companies and their clients, subcontractors and stakeholders to produce better projects. This report has identified a range of indicators against which companies can measure their performance and contribution to more sustainable construction. Indicators are powerful tools but are ultimately ineffective unless companies compile, analyse and act upon the information they provide. This section identifies the proposed next steps for companies wishing to make use of the indicators.

5.3.1 Issue identification

To select appropriate indicators, the first stage for any company is to identify those sustainable construction issues that are of greatest relevance to its business. Not all the sustainability issues addressed in this report are relevant to all companies or projects. For example, energy efficiency may be of prime importance for the design and occupation of housing projects, but is likely to be less relevant for a bridge or flood defence scheme. Furthermore, there may be important sustainability issues that are not addressed. This will involve the company considering its principal environmental and social impacts through review of its business activities and matching these against the issues identified within this report.

5.3.2 Indicator identification and development

Once the company has assessed the principal sustainability issues and ways in which it can most effectively contribute to more sustainable construction, the second stage is to identify indicators that can be used to assess current and future performance. This report identifies relevant indicators for key sustainability issues. The list is not exhaustive and companies may wish to develop their own measures.

It is hoped that most companies will choose to measure and report progress against the strategic company indicators presented in this report as most indicators are relevant to most businesses. Companies should also identify those operational indicators that are relevant to their business or to projects for which performance is being measured.

5.3.3 Establish data compilation procedures

Once the indicators have been agreed the next step is to establish cost effective data compilation procedures to compile the required information and identify one or more individuals responsible for compiling the data.

The indicators proposed in this report do not include prescribed methodologies to compile the data. The process of developing an appropriate methodology to measure performance for specific projects will help the company to install efficient procedures for data compilation and strategies to improve performance. It is important that procedures are transparent in order that data is compiled on a consistent basis, especially across projects and among different project managers.

Case study 8 ❍ Page 73

...describes the process by which Carillion have developed a KPI for energy use in construction, the issues and difficulties involved in compiling the required data, and their pragmatic solution.

Compilation of the data should be undertaken as part of routine project or environmental management and quality systems to ensure the required data is always produced efficiently and accurately. Developing effective and efficient systems and procedures for data compilation is crucial to successful implementation of operational indicators.

Industry organisations, or major Government clients, have a key role in helping companies working within the same sub-sector of the industry to use similar indicators; for example, housebuilding or road construction. These groups could also usefully develop consistent methodologies to facilitate benchmarking of performance between companies.

5.3.4 Data compilation

Once procedures to compile the information have been established these should be implemented together with a programme to ensure that they can be updated and progress assessed on a regular basis. For most indicators this is likely to be quarterly. Suppliers will also need to be informed about the indicators prior to requests for appropriate data.

Most of the operational indicators require information to be compiled at the project level and collated to give an aggregate company value. To be effective, an individual that is also able to influence the project performance should collate the project level information.

Case study 9 ❍ Page 74

...describes Balfour Beatty's efforts to develop operational indicators and the challenges of establishing efficient procedures for data compilation.

Training will be a key element in the programme to roll out use of the indicators. Training should also be used to raise awareness of the key sustainable construction issues. The focus should therefore be on how project performance can be improved; use of indicators to evaluate progress should be a secondary, but important consideration.

For large companies, working on many projects, it may not be cost-effective to collate information from every project. From a practical perspective large companies may therefore choose initially to select larger projects, or those from specific sectors of the business, to trial and refine both the indicators and their approach to data collation. Some of the proposed operational indicators measure performance normalised against the total number of projects undertaken by the company or business turnover. Where companies are not recording information from all the projects they undertake the performance would need to be normalised against the proportion of total projects for which data has been compiled.

5.3.5 On-going assessment and indicator development

It is likely to take about a year to compile a robust baseline of data against the strategic indicators or from a reasonable cross section of projects. Once this has been achieved the data can be used to determine the current position and identify targets for improvement. Where appropriate, the processes by which the indicator data are collated and compiled should also be modified to reflect the initial experience and to improve the robustness of the measurement.

Once data from a representative cross-section of the company projects have been acquired these can be collated to produce and aggregate company average and range of values for different projects. This collation and analysis of data provides the means through which the company is able to evaluate performance and those of individual teams and managers.

Companies are likely to find it beneficial to work with competitors and related businesses within benchmarking clubs to learn from others and develop a consistent basis for measuring performance. Through benchmarking clubs, companies will also be able to compare their own performance against others, to stimulate competition and the drive for continuous improvement.

5.3.6 Improving performance

Once a robust baseline of current performance against the indicator is available, the company should consider what strategies and plans are needed to make progress and what target would be appropriate against which to assess this progress. A range of general guidance is available on overall sustainability performance improvement and specific aspects of environmental and social performance (CBPP, 2000).

The ultimate intention of the operational indicators is to improve project performance. Whilst the data collection activities required by the indicators will constantly remind project managers of their responsibility for more sustainable construction it must be supported by training and management plans and revised procedures to improve project delivery. As part of the ongoing data compilation staff should therefore receive regular training on how to improve project performance as well as on how to calculate performance measures.

5.3.7 On-going monitoring and reporting

Once a baseline of data has been prepared and strategies and targets developed to advance and measure progress the company should consider reporting progress on a regular basis. Reporting should be both internal – to employees to demonstrate the importance of maintaining progress – and external through development of environmental and social reports. Ideally such reports should be externally validated.

5.3.8 Involvement in industry initiatives

A further way companies can increase their knowledge of and contribute to delivery of more sustainable construction is to become involved in industry initiatives. In the last two years, encouraged by the Government, the industry has implemented many initiatives to raise awareness and improve the performance of the industry, including these:

- the Sustainable Construction Focus Group that was established and has published *Towards Sustainability – A Strategy for the Construction Industry* (SCFG, 2000)
- the Construction Best Practice Programme has developed 10 Key Performance Indicators to measure progress against the targets established in *Rethinking Construction (DETR, 1998)*
- M4I has established several demonstration projects with important sustainability elements and is developing a measure of sustainable project performance
- the Construction Best Practice Programme, CIRIA and others have published guidance to improve awareness
- the Construction Industry Environmental Forum has held numerous seminars to raise awareness and improve knowledge of sustainability issues
- numerous sustainability working groups have been established by M4I, the Housing Forum, the Construction Industry Board, trade associations and others
- benchmarking clubs have been established (CBPP, 2000b).

Involvement by companies in these activities is an excellent means of sharing information and ideas to improve company performance.

5.4 COMPANY CHECK-LIST

Table 5.1 provides a checklist to assist companies develop and benefit from use of sustainable construction indicator

Table 5.1 *Checklist of company actions*

- ➤ identify a director with overall responsibility for promoting sustainable construction and use of the indicators and appoint a person (or team) to oversee the development and introduction of the indicators. Allocate appropriate resources ❑
- ➤ review the list of strategic indicators and establish procedures to compile the necessary information for each ❑
- ➤ review the list of operational indicators in consultation with employees, clients, major suppliers and other relevant stakeholder groups. Identify those operational indicators of greatest relevance to the company and/or develop more relevant measures of overall company project performance ❑
- ➤ discuss the use of operational indicators within sectoral organisations with the aim of agreeing a common suite of indicators with competitors in order to benchmark performance ❑
- ➤ review company records to assess whether existing data can be used to develop a baseline against which to measure current performance ❑
- ➤ publicise and promote the indicators and their objectives with employees and stakeholders and provide training to key staff in how to deliver more sustainable construction projects ❑
- ➤ establish targets for each of the selected indicators in consultation with stakeholders and staff ❑
- ➤ amend company procedures that encourage unsustainable practices and inform and train staff in the new and improved practices ❑
- ➤ implement procedures for consistent, time effective data gathering by developing simple tools and incorporating data gathering within quality procedures ❑
- ➤ initiate a general staff training programme to promote best-practice and deliver the improvements that are sought ❑
- ➤ monitor data gathering and performance against each of the indicators. Amend the procedures for data gathering and the indicators with increasing experience ❑
- ➤ routinely evaluate the performance of the indicators against the targets that were set and implement strategic plans and further training where performance is identified as falling behind the intended rate of progress ❑
- ➤ after a year of consistent data gathering publicise and report performance against the strategic indicators (and selected operational indicators) within an independently audited company sustainability report ❑
- ➤ continue to monitor performance on an annual basis against the indicators seeking, wherever possible to maintain a consistent data record ❑
- ➤ encourage and promote the use of indicators within client, supplier and competitor organisations ❑
- ➤ get involved in industry initiatives to promote more sustainable construction ❑
- ➤ review the targets and indicators in use and revise and extend these as appropriate. ❑

6 References

BiE 2000, *Business in the Environment* – Annual Company Survey 2000

BRE 2000, Building Research Establishment, *Index of Project Sustainability* (complete)

BSRIA 2000, Building Services Research and Information Association, www.bsria.co.uk/bsriaweb/research/current/environment/default.htm

CBPP 2000 *KPI Pack, Construction Best Practice Programme*

CBPP 2000, *Construction Best Practice Programme*, www.cbpp.org.uk

CBPP 2000a, *The Construction Benchmarking Gateway*, www.KPIzone.com

CC 2000, *A Framework for Business Improvement, Investors in People and the Construction Industry*, Construction Confederation

CCS 2000, Considerate Constructors Scheme

CIB 2000, Construction Industry Board, Considerate Constructors Scheme, www.ciboard.org.uk

CITB 1999, *Construction Employment and Training Forecast 2000–2004*, Construction Industry Training Board

CIRIA 1999 *Environmental Issues in Construction – a Strategic Review*, Construction Industry Research and Information Association

CIRIA 2000 *The handbook of supply chain management*, Construction Industry Research and Information Association

CSD 2000, *Sustainable Construction – the data*, CR258/99, Centre for Sustainable Construction, Building Research Establishment

DETR 1998, *Rethinking Construction*, The Report of the Construction Task Force, Department of the Environment, Transport and the Regions

DETR 1999, *A better quality of life – a strategy for sustainable development for the UK*, Department of the Environment, Transport and the Regions

DETR 1999a, *Quality of life counts, Indicators for a Strategy for Sustainable Development for the UK: a Baseline Assessment*, Government Statistical Service

DETR 1999b, *Guidelines for company greenhouse gas emissions reporting*, www.detr.gov.uk/environment/envrp/gas/index.htm

DETR 1999c, *Towards an Urban Renaissance. Final Report of the Urban Task Force*. Department of the Environment, Transport and the Regions, London

DETR 2000, *Building a better quality of life, a strategy for more sustainable construction*, Department of the Environment, Transport and the Regions

DETR 2000a *KPI Report for the Minister of Construction*, Department of the Environment, Transport and the Regions

DETR 2000b, *Environmental Reporting*, www.environment.detr.gov.uk/envrp

EC 1996, *Environmental Impact Assessment, Guidance on Scoping*, European Commission DGXI

Environment Agency 1999, Speech by Ed Gallagher to the Construction Confederation Environment Conference, March 1999

GCCP 2000, *Achieving sustainability in construction procurement,* Government Construction Clients' Panel, June 2000, www.hm-treasury.gov.uk/gccp

IIP 2000, Investors in People, www.iipuk.co.uk

M4I 2000, *A Commitment to People "Our Biggest Asset"*, a report from the Movement for Innovation's working group on Respect for People

NEF 1998, *Participation Works!*, NEF/UK Participation Network, New Economics Foundation

NEF 2000, New Economics Foundation, www.neweconomics.org.uk

RIBA 2000, RIBA Practice Services, *Architects and the Changing Construction Industry*, www.site.yahoo.net/practice-services/conchan.html

SCFG 2000, *Towards Sustainability – A Strategy for the Construction Industry*, Sustainable Construction Focus Group, Department of the Environment, Transport and the Regions

A1 Sustainable construction key themes

A1.1 Avoiding pollution

Polluting the environment generates negative publicity and additional costs associated with fines or necessary remediation works. It may also mean that regulators and planning authorities seek additional controls on or raise more objections to future company operations. Being identified as a polluting company also sends a clear signal to potential clients or employees that the business is not effectively managed. Avoiding pollution requires construction workers to understand the potential impacts of activities upon the local environment and companies to implement appropriate environmental management systems. The benefits of effective environmental management systems are also seen in more efficient working practices.

A key element in avoiding pollution is waste minimisation, in which a clear hierarchy of preferred options exists:

- reduce generation of waste
- re-use waste for either the same or a different purpose
- re-cycle through composting or energy recovery
- dispose using the best-practicable environmental option.

A1.2 Protecting and enhancing bio-diversity

The potential for long-term degradation to the environment through construction activities is considerable, in particular by quarrying, unsustainable forestry and new building on greenfield sites. Good site management and design can mitigate or minimise harm but the construction industry is also offered tremendous opportunities to enhance the local environment, through urban regeneration, good design and landscaping resulting in benefits such as habitat creation and the production of sensitive ecosystems. Enhancing bio-diversity improves the local environment and is good for public relations, particularly with local residents, and may facilitate the securing of required planning consents.

A1.3 Efficient use of resources

More sustainable construction requires clients, designers and those involved in the construction process, to select and source materials to minimise the environmental impacts. For sustainable sourcing of materials there are a number of preferred options:

- refurbishment and renovation of existing assets is preferred to new build.
- recycled materials are preferred to non-renewable products, which where appropriate should be from sustainable supplies.
- prefabricated materials are preferred to those constructed on site.
- local products are preferred to those transported over long distances.
- renewable materials and those sourced from sustainable managed supplies are preferred.
- products with the lowest embodied or lifetime energy are best.

The choice of materials to use in construction will always be a balance between conflicting requirements but in many instances it is cheaper to re-use materials on site, use local products or choose one with a low embodied energy. The priority is to ensure the preferred option is used whenever it is appropriate, and always seriously considered within the design and construction process.

Water is another important resource that can be conserved through good design including application of greywater collection and low flush toilets. Efficient construction practices can also minimise water use during the building phase.

A1.4 Improved energy efficiency

The need to improve the energy efficiency of buildings, particularly domestic housing, is a priority of the UK Climate Change Programme. For occupants it also makes good financial sense. Energy consumed by buildings in use makes the overwhelming majority of the total energy consumed during the structure's lifetime, and design and construction of energy efficient buildings delivers the greatest returns. This can be achieved by techniques including better insulation and use of passive systems for lighting, air movement and cooling such as thermal mass.

In addition to design and construction of energy efficient buildings, the construction industry can deliver more energy-efficient structures by:

- making greater use of renewable energy sources or combined heat and power schemes
- minimising energy consumed during construction processes
- reducing energy consumed in its own offices and practices.

Combined heat and power schemes and renewable systems such as solar power reduce the environmental impacts associated with electricity generation. To reduce energy consumed during construction, consideration should be given to the materials used and where they have been sourced from (as described above). Use of local labour also reduces travel costs, helps the local community and is easier for employees. In conducting their own business companies can seek to use more energy efficient buildings, reduce their own travel or introduce green travel plans.

A1.5 Respect for staff

Improving the delivery of projects, and sustainability of built assets, relies upon improving the knowledge, skills and performance of employees working within the industry and optimising the performance of project teams. This is achieved by providing:

- effective management
- equal opportunities
- a safe and conducive environment for work
- opportunities for training
- flexible working conditions and arrangements.

Effective management recognises and responds to the strengths and limitations of each employee and motivates them accordingly. Contented, motivated staff are more productive. They are also less likely to leave, reducing the inevitable disruption to business activities and recruitment and training costs and helping to maintain institutional knowledge.

Retaining skilled staff is crucial to all businesses. The construction industry has an ageing workforce and serious skill shortages in several disciplines (DETR, 2000). The industry therefore urgently needs to attract and retain more of the most able individuals, and to ensure that the pool from which it is drawing prospective employees is as wide as possible. To do this the industry must be more attractive to women and people from ethnic minorities. One way of encouraging more women into the industry is greater use of more flexible working arrangements. Employers must also ensure that all employees have equal opportunities so that once recruited they can be retained within the industry or business in which they are employed.

In addition to recruiting and retaining able staff, training to update and improve skills to improve productivity and performance are essential to any business. Trained staff will be safer, more motivated and produce higher quality work, all of which significantly contribute to the business. Delivering planned and structured training for employees and for subcontractors engaged on long-term assignments is therefore essential to produce sustained improvements in business performance.

Training is particularly important to improve the sustainability of construction since this requires employees to respond to a much wider range of issues and considerations than the traditional construction process.

One of the key elements of more sustainable construction is providing a safe and conducive working environment for employees. Construction is an inherently dangerous industry. In 1999, 48 construction workers were killed on site and 13,000 injured. However, risks on site can be minimised by sensible design and implementation of good working practices and comprehensive training. Safety at work however goes beyond avoidance of injury or death and includes the prevention of harassment, intimidation or bullying of staff, linked to provision of equal opportunities for all.

Maximising employee performance also requires a conducive working environment that reflects the needs of employees. For the construction industry, providing adequate services for workers on site is particularly important to help motivate project teams and retain continuity by avoiding staff losses.

A1.6 Working with local communities

An economy based on people's needs must allow everyone to contribute to the decisions affecting them (NEF, 2000). Working with the local community to maximise the benefits arising from the built asset, whilst seeking to mitigate or minimise negative impacts, is a key element of more sustainable construction. It is also one that requires all sectors of the industry to work together in a coherent and consistent manner. The construction industry impacts upon local communities through::

- the actual physical structures built, which may enhance or degrade the local environment and economy
- activities during occupation and use
- the construction process that generates employment prospects but can also be disruptive, noisy and dirty
- delivery of materials to site.

Local councils, through the planning process, clearly have the leading role in deciding whether a proposed development can proceed. However, working with local communities from conception of the proposed project can only help to obtain the required permissions. Local communities include three main groups:

- **local neighbours:**
  - residents
  - retailers
  - schools
  - businesses and their employees
- **local interest groups:**
  - local NGOs
  - Residents' Associations
  - Chambers of Commerce
- **wider community**
  - drivers moving through the area
  - national NGOs.

For many projects it will be necessary to engage with all of these stakeholders and create different channels of communication with each, reflecting their different needs and interests. It is particularly important to involve groups that are often marginalised including children, people from minority ethnic communities and older people.

Local consultation and preferably participation generates excitement, energy and a sense of ownership, providing local expertise to the project, legitimacy and enhanced quality of life, and helping to build more sustainable communities. In seeking to engage with the local community a range of proven techniques have been developed (NEF, 1988).

Once planning permission has been secured it is possible for new built assets to be imposed without discussion upon an unwilling local community. However, proceeding in this manner will escalate the costs and may delay the completion of the project. In particular, by listening to local community concerns it may be possible to:

- alter the development so that it more fully meets local needs and thus avoid costly delays through planning appeals and inquiries
- prevent direct action against the site, reducing costs of security and delays to the project through removing objectors
- add local value through the project
- enhance the reputations of all those involved in the project with subsequent indirect benefits.

In design of the structure, landscaping can help to ensure that the structure is sympathetic to the local environment and that green areas, as far as possible, are preserved or enhanced. Clients and consultants can also ensure transport to the site is adequate and will not ıead to excessive congestion upon local roads. During the construction process contractors can ensure disruption to the local community is minimised, such as by operating the site under Considerate Constructor and similar schemes (CIB,2000). Underlying the process, clients must also recognise that seeking to address local community concerns is an important element of new construction projects and that the scheme will need to reflect this.

A1.7 Partnership working

The previous sections identified employees and local communities as important stakeholders in delivering more sustainable construction. The third key stakeholder group is business partners, with whom new types of relationships and ways of working need to be developed to move the industry away from traditional adversarial approaches and towards greater sustainability.

Rethinking Construction (DETR, 1998) proposed that the industry move away from competitive tendering towards developing long-term relationships throughout the supply chain based on clear measurement of performance and sustained improvements in quality and efficiency. Partnering arrangements involve two or more organisations working together to improve performance through agreeing mutual objectives, devising a way for resolving any disputes and committing themselves to continuous improvement, measuring progress and sharing the gains (DETR, 1998). Partnering arrangements are inherently more sustainable than the conventional staged approach to construction. Increasing the number of projects and scope of work undertaken using this model is therefore a key theme for more sustainable construction. Cost savings of up to 50 per cent have been reported for series of projects undertaken under partnering arrangements and time savings of up to 80 per cent (DETR, 1998; CIRIA, 2000).

Conventional construction involves delivery of a built asset through competitive tendering against specifications, with the aim of achieving the lowest price for each element of the work rather than the best value for the overall project. Sustainable construction seeks to optimise capital expenditure over the lifetime of the asset.

Partnering arrangements reduce barriers to using the skills and knowledge of suppliers and constructors effectively in the design and planning of the projects. They also enable teams to learn from past experience and achieve incremental improvements in subsequent performance and to innovate more effectively. Stable teams will also have a greater stake in the final outcome leading to improved product delivery.

Partnering arrangements can be extended throughout the supply chain to develop integrated project processes which avoid the contractual and confrontational culture that still persists in some parts of the industry. This presents major challenges to constructors and suppliers and will require partners to have an opportunity to share in the rewards of improved performance as well as the risks. Clients will need to rethink contractual relationships; and all need to emphasise the importance of change in the industry in order to deliver more sustainable construction and the benefits that could accrue to all.

A1.8 Business profitability and performance

The need to improve the profitability and performance of the industry is another key element of more sustainable construction. Rethinking Construction (DETR 1998) described an under-achieving industry poorly perceived by many of its clients and investors (see Section 1.2). It highlighted a cycle of unsatisfactory project delivery and under-investment linked to low profitability arising from an overemphasis by clients upon price instead of value (Figure A.1).

Figure A.1 *Cycle of under-achievement*

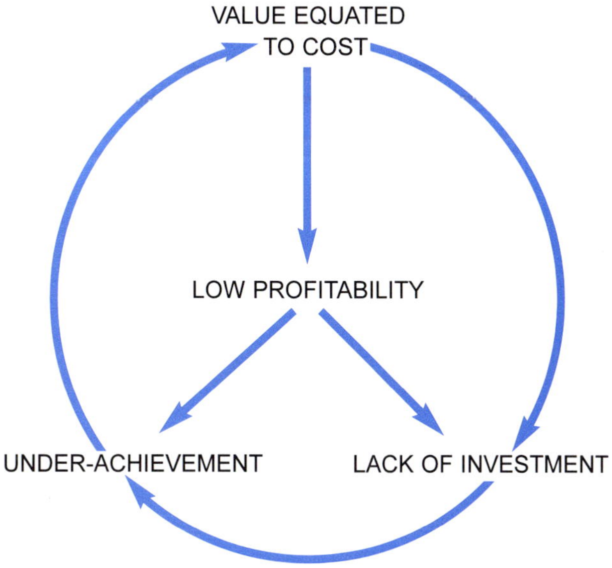

Too many clients still equate value with cost, and as a consequence profitability rates are unpredictable and low. Tight margins on projects have in turn contributed to a perception of unsatisfactory project performance by clients. Tight margins have also contributed to low and unpredictable company profitability, which in turn has led to reduced expenditure on research and development, so damaging the industry's capacity to innovate and train staff. There is also under-investment of capital. Breaking this cycle of business under-performance is essential, if the industry is to innovate and achieve the benefits that will arise from more sustainable construction.

Low and unpredictable profitability has significantly contributed to a lower valuation for construction companies and poor share price performance relative to FTSE indexes. Improving profitability is key to developing long-term relationships with investors and enhancing investment and capitalisation.

A1.9 Improved project delivery

Better project delivery is also crucial in order to build the trust needed to promote partnership working. The proposals for the industry within Rethinking Construction therefore address the economic aspects of more sustainable construction and also touch upon those contained within Respect for Staff (health and safety).

The Rethinking Construction Task Force identified five key drivers of change:

- committed leadership
- a focus on the customer
- integrated processes and teams
- a quality driven agenda
- commitment to people.

They proposed ambitious targets and challenged the industry to measure its performance against these. To provide a consistent basis against which companies and the industry could measure its performance the Key Performance Indicator (KPI) Working Group was established, reporting in January 2000 (DETR, 2000a). The KPIs have been developed by the Construction Best Practice Programme (CBPP, 2000) to provide a consistent basis against which to measure business profitability and productivity.

A1.10 Monitoring and reporting of performance against targets

The tenth Key Theme of the DETR Sustainable Construction Strategy is the need to monitor and report performance against targets. The multidimensional and sometimes conflicting objectives of sustainable construction make measuring the progress of the industry or individual businesses essential. This is both to monitor the effectiveness of policies and initiatives and to stimulate change by setting targets and working towards these. Indicators and targets are identified within the DETR Strategy (DETR, 2000) as an essential mechanism for measuring sustainability performance and hence delivering more sustainable construction.

A2 Framework for sustainable construction indicators

Delivering sustainable construction needs action at a variety of levels and by participants at each stage in the life-cycle of a construction project. In order to support this action, indicators are needed to measure performance for each of these levels and at all stages of the construction life-cycle. Section 1.3.2 introduced some current industry initiatives required to develop this comprehensive suite of indicators. This Appendix provides further information to enable companies to become involved in these initiatives where appropriate.

Due to the diversity of the construction industry is not possible to develop a single comprehensive, coherent suite of indicators relevant to the entire industry or companies within it. Separate suites of indicators, and related targets, are therefore needed to stimulate and monitor change throughout the industry. The CIB had agreed to co-ordinate the preparation of an indicator framework to ensure that initiatives at all levels (project, company and industry) and across all sectors (clients, architects, material producers and others) are consistent and simple to use. Figure A.2.1 illustrates the framework of indicator suites currently under development.

This report has focused on the use of company indicators, specifically for companies that provide design and construction services. However, the research, consultation and industry testing that underpin the report (section 1.3.2) also covered project and industry-level indicators.

A.2.1 PROJECT INDICATORS

At the project level, sustainable construction involves a skilled, motivated and safe workforce delivering successful, profitable projects that enhance the quality of life of users and local residents. Projects should also be constructed in a manner that consumes the least possible resources and minimises pollution and nuisance. Measuring the sustainability of individual projects is therefore a key element of monitoring progress towards more sustainable construction. Examples of project indicators are:

- tonnes of waste arising to landfill per £ construction value
- average distance (km) travelled to site by workers.

The Movement for Innovation (M4I) is the leading initiative promoting project performance improvement for non-housing construction in the private sector. Two working groups have been established to promote improved sustainability performance:

- respect for people working group – covering recruitment, retention and respect for staff (M4I, 2000)
- sustainability working group – covering environmental performance.

Both groups have developed indicators to measure performance and will use the indicators to determine performance of the industry at large and specifically to measure the business and organisational benefits from the M4I Demonstration Projects.

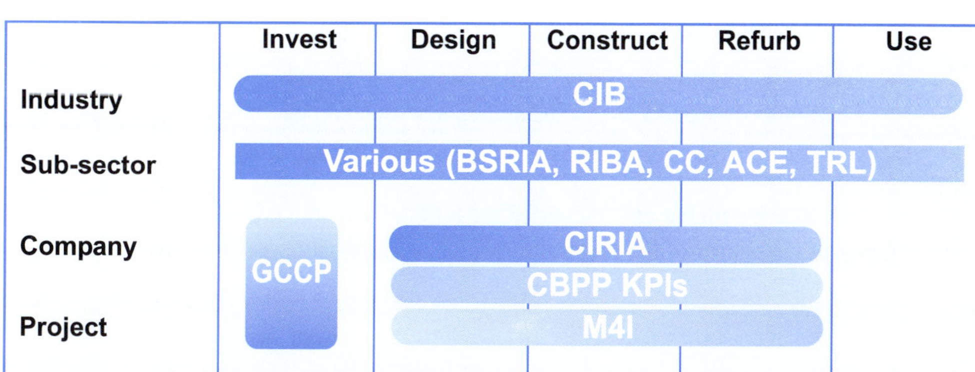

Figure A 2.1 *Suites of sustainable construction indicators*

The Sustainability Working Group has been developing a suite of Environmental Performance Indicators (EPIs) for use by clients, designers and contractors on building and construction projects. These EPIs are intended to facilitate the planning and design of more sustainable construction outcomes, in terms of the performance of the built product, the impact of the construction process on the environment, and ongoing measurement of environmental performance post-occupancy.

This CIRIA report uses project indicators to develop company operational indicators (Section 1.3.5), based upon aggregating project-level information. It is important therefore, in choosing indicators of project performance, that they can be readily aggregated to give meaningful information at the company level. The report identifies a wide range of operational indicators that can be used to inform the development of the M4I indicators.

A.2.2 COMPANY INDICATORS

Sections 2-4 of this report propose a suite of Company indicators that can be used by businesses providing design or construction services in order to measure their sustainability. Section 5 considers in detail the benefits of sustainability reporting and benchmarking of company performance and the role of targets. Section 5 also describes an action plan for companies wishing to develop and use indicators.

A.2.3 SECTORAL INDICATORS

Different sustainable construction indicators and targets are needed for specific sub-sectors of the construction industry such as road building and house building. Development of appropriate indicator suites enables performance trends within the sector to be monitored, performances to be compared, and key areas of concern to be identified.

A number of organisations are currently engaged in developing indicators relevant to particular sub-sectors including:

- the Building Services Research and Information Association (BSRIA) producing indicators for buildings in use (BSRIA, 2000)
- the Royal Institute of British Architects and Construction Industry Council are developing a suite of design quality indicators (RIBA, 2000)
- TRL is working on measures of performance for road construction and maintenance
- the Association of Consulting Engineers is looking at company indicators against which their members can benchmark performance.

A.2.4 INDUSTRY INDICATORS

Within *Building a Better Quality of Life* (DETR, 2000) the Government expressed its commitment to tracking the progress of the industry towards sustainable construction through a suite of indicators. Some indicators within Quality of Life Counts (DETR, 1999a), presenting measures of the sustainability of the UK, are already directly relevant to the construction industry. These include:

- construction and demolition waste going to landfill
- primary aggregates used per unit of construction value
- amount of secondary/recycled aggregates used compared to virgin aggregate
- carbon dioxide emissions by end user.

The Government has also identified the KPIs as possible project indicators that could be aggregated to provide company and sectoral measures of sustainable construction.

The challenge of developing a comprehensive suite of industry-level sustainable construction indicators has been taken on by the Construction Industry Board working with DETR to collect performance data for the whole construction industry. Social and environmental indicators will be added to the national set of Key Performance Indicators (KPIs) and promotion will be co-ordinated with the existing KPIs.

Table A.2.1 lists possible industry-level indicators that could be considered as part of the CIB initiative. These were developed during the CIRIA research project, following a review of relevant indicators that are currently in use or could be readily derived from existing data sources. The list is not definitive but provides an initial assessment of measures that could be used to assess the current status of the industry and future progress on each of the key sustainability themes identified within the report.

Table A 2 *Preliminary list of possible industry-level indicators*

Theme	INDICATOR
Avoiding pollution	environmental incidents arising from construction activities per £1m construction output (Incidents based upon data from the Environment Agency, construction output from the DETR)Number of construction companies certified to ISO14001 or equivalent standard (BSI) number of sites operating under the Considerate Constructors Scheme per £1m construction output *(Considerate Constructors Scheme, construction output from the DETR)* value of environmental enhancement activities as a proportion of total construction output (under development by English Nature) inert waste disposed of to landfill per £1m construction output (based upon *HM Customs and Excise for waste, construction output from the DETR)*
Protecting and enhancing biodiversity	number of specially designated features destroyed, partially destroyed or unfavourably declining due to development activities *(English Nature)* area of greenfield land used for construction purposes per £1m construction output (km^2) *(DETR)*
Efficient use of resources	annual water consumption in buildings per person (m^3) *(BRE)* primary aggregates used per £1m construction output *(DETR)* percentage of timber sourced from renewable supplies (to be developed)
Improved energy efficiency	carbon dioxide emissions from domestic and commercial sources per m^2 of built floorspace *(emissions for DETR, floorspace statistics from BRE)* total embodied energy and carbon dioxide emissions per £1m construction output (to be developed) Carbon dioxide emissions from transport of construction materials (to be developed)

Continued on next page

Theme	INDICATOR
Respect for staff	number of construction companies registered as Investors in People or equivalent qualification *(Investors in People)* net surplus or deficit of skilled labour *(Construction Industry Training Board)* days lost through stoppages in the construction industry *(Digest of Data for the Construction Industry)* number of staff entering the industry with S/NVQ level II or III *(Construction Industry Training Board)* number of women and people from ethnic minorities entering the industry with S/NVQ level II or III *(Construction Industry Training Board)* reportable fatal and non-fatal accidents per 100,000 hours worked *(HSE)*
Working with local communities	number of projects that implement a plan for stakeholder dialogue *(Survey)* number of independently audited environmental or social responsibility reports published by construction companies *(Survey)*
Partnership working	proportion (by value) of construction projects for which whole-life costs have made the basis of design and construction decisions (to be developed)
Improved project delivery	international productivity comparison for construction (Digest of Data for the Construction Industry)
Increased productivity and profitability	output of the construction industry as a percentage of GDP *(Digest of Data for the Construction Industry)* construction manpower - seasonally adjusted (number of people) *(Digest of Data for the Construction Industry)* value of construction new orders by contractors (£) *(Digest of Data for the Construction Industry)* insolvencies and bankruptcies in the construction industry *(Digest of Data for the Construction Industry)*

The proposed indicators are largely based upon existing national datasets. Use of existing data sets enables trends in the indicators to be produced using historic data that is important in assessing the direction and rate of progress. Accordingly, most industry-level indicators will need to be compiled using a top-down approach. Where appropriate national data sets are unavailable, national data can be obtained by aggregating company indicators to provide the required information. This requires companies to be surveyed on an annual basis and is relatively expensive.

Representative results from surveys will only be obtained when the company has the necessary information. For many of the sustainable construction themes this is unlikely at the present time. The diversity of the companies involved in the construction industry makes undertaking representative surveys particularly difficult and adds to their cost. That said, to compile a comprehensive suite of indicators, surveys are likely to be the only option available to obtain the necessary information.

A3 Examples of targets in use

Table A.3 *Examples of targets in use*

Issue	Target	Source
Waste minimisation and management	to reduce the amount of construction and demolition waste going to landfill to 85 per cent of 1998 levels by 2005 (Northern Ireland)	indicators for a strategy for sustainable development in the United Kingdom: a baseline assessment, DETR, 1999
Mitigation and management of pollution from site	75% participation in ISO 14001 and/or EMAS among FTSE 100 companies (ie at least one certified or registered site) by 2001	indicators for a strategy for sustainable development in the United Kingdom: a baseline assessment, DETR, 1999
Optimising use of brownfield sites	60% of new homes and conversions to be on previously developed land by 2008 in England	a better quality of life, A strategy for sustainable development in the UK planning for communities of the future, DETR, 1998
Designing for whole-life costs	100 per cent of departments to use 'whole-life cost' assessment as part of the technical appraisal for decision to build new or refurbish/re-use existing assets by March 2003 100 per cent of departments to have criteria for evaluation of sustainability included in procurement procedures for construction projects by March 2003	constructing the Best Government Client; Achieving sustainability in construction procurement, Sustainability Action Plan, The Government Construction Clients' Panel (GCCP) Sustainable Construction Action Plan
Water conservation	reduce water in offices and depots to 30 per cent below accepted norm for office type or 1996/7 consumption whichever is higher	Environment Agency Environmental Report 1998/9; Environmental Performance Targets

CASE STUDY 1

The Arup Campus – sustainable office design through objective target setting

ARUP

The new Arup Campus brings together two major offices on a single UK site that once completed will support 650 staff.

The specific brief given to Arup Associates' in-house design team was to achieve a balanced resolution between environmentally sustainable design objectives and commercial realism. Accordingly a "long-life, loose fit, low energy design" process was adopted while still creating a comfortable, healthy, well-equipped work and social environment.

From the concept stage a multi-disciplinary team of environmental, R&D and building expertise established environmental objectives and targets for the Campus. These targets encouraged the development of an innovative, environmentally sensitive design that included:

- natural ventilation in most areas
- optimal use of natural lighting, whilst minimising glare and solar gain
- minimising energy consumption
- functional use of natural materials
- use of industry best practice including use of BREEAM 98 to set targets.

Arup Campus

In most cases it proved possible to improve upon the targets set during the design process to achieve higher levels of performance as shown in the table below.

Compared to the existing two offices the new Arup Campus also demonstrates a significantly enhanced performance in terms of eco-efficiency, minimising energy use and use of natural materials, including a 40 per cent reduction in electricity consumption and 50 per cent reduction in gas consumption.

	Design target and actual performance	Principal methods
Reduced water consumption and use	performance of 15 litres/person day was achieved against a target of 18 litres/person day	use of toilets with six litre double flush facility; manually controlled temperature mixed flow taps with spray nozzle in hand basins; shower flow rate of 0.1 litres/s
Reduced energy consumption for lighting	average annual electrical load of 8w/m^2 with supporting target of an average of 5 per cent daylight factor	a combination of large roof pods to introduce daylight and efficiently located, dimmed luminaries. Solar glare was reduced by manually and/or electrically operated blinds with external motorised blinds and shutters where required
CO_2 emissions per year	targets were established of 50 kg/m^2 overall and 31/70 kg/m^2 for non/air-conditioned areas. An average of 28 kg/m^2 has been achieved	use of high efficiency condensing boilers with a management system that resets required boiler flow temperature to achieve maximum efficiency and maximum condensation

CASE STUDY 2

Development of company indicators for design and construction companies

CIRIA Research Project 609 has developed sustainability indicators for companies providing design and construction services. These indicators have been developed following extensive research and consultation with the industry and its stakeholders.

Initially, a preliminary suite of indicators was developed, drawing upon previous research. This preliminary work identified a large number of environmental and economic indicators but revealed that knowledge of social issues in the construction industry was much less well developed. A significant amount of new research and consultation was therefore undertaken to identify social issues and potential indicators. This activity was led by the New Economics Foundation, a respected sustainable development think tank.

The preliminary suite of indicators was tested by these project partner companies:

- Carillion Plc
- Ove Arup Partnership
- Balfour Beatty
- Bovis Lend Lease
- Buro Happold
- John Laing
- Scott Wilson
- Montgomery Watson
- Taylor Woodrow Construction
- WSP Group.

The testing process involved an initial review of the provisional indicators followed by selected data gathering to test the feasibility of the proposed measures. The pilot companies tested the relevance, appropriateness, ease of compilation and value of information generated for each of the proposed indicators.

> **Delivery of sustainable construction puts design right at the heart of the agenda, especially since decisions made early in the design process can be the most significant influence on a project's ultimate sustainability. These indicators provide an effective map to show design and contracting organisations what factors they should jointly address to improve their own sustainability performance. Clearly this includes those areas where they should be attempting to influence client decisions by setting some performance targets.**
>
> **Michael Dickson**
> **Chairman, Buro Happold**
> **Chairman, Construction Industry Council**

Feedback from the piloting was used to improve the practicality of the indicators and the revised suites were presented in a consultation paper issued in May 2000 to more than 500 industry and stakeholder representatives. A consultation workshop in June, attended by representatives of companies from across the industry, planning and regulatory authorities, Government Departments and agencies and other stakeholder groups provided a further opportunity to comment on and improve the proposed measures. About 100 organisations responded to the consultation and their comments have been used to further refine the indicator suites.

The final stage in developing the company indicators has involved extensive review and comment by members of the project steering group and team. The final suite of indicators, and how companies can use these, are presented in this report.

CASE STUDY 3

MONTGOMERY WATSON

Montgomery Watson sustainability KPIs

Montgomery Watson (MW) maintains a series of Key Performance Indicators (KPIs) measuring the technical and financial performance of projects and the company. As part of the testing phase of the CIRIA RP609 project MW undertook an in-depth evaluation of selected sustainable construction KPIs for company use. The evaluation of sustainability KPIs utilised the knowledge of in-house specialists and other staff in each field to assess how the KPIs could be applied in practice.

Process

The evaluation process asked staff to rank as Low/Easy, Medium or High/ Difficult a list of KPIs according to their effectiveness and their cost-effectiveness. Effectiveness was defined as 'What will drive MW and our clients and projects, most effectively towards sustainability?'

The Cost-effectiveness of the KPI was evaluated in terms of how 'do-able' it was to implement and measure. 'Easy' meant that the data was readily available; 'Medium' meant available, but no system in place to collect it; and 'Difficult' no data available and no system.

Some indicators showed a wide range of voting, particularly on effectiveness. Analysis of differences led to some illuminating discussion and helped us move towards a common interpretation of the KPIs.

MW staff brainstorm the effectiveness of proposed indicators

Results

The process enabled MW to prepare a shortlist of indicators for adoption and provided feedback to CIRIA on the viability of the indicators. It also produced these conslusions:

- The introduction of indicators should initially focus on those that are highly effective and not too difficult to implement before developing procedures to measure highly effective indicators that are not cost-effective using current practice for data acquisition.
- Normalisation of data should be carried out consistently, eg using units of output or production. Measures of output or production tend to be sector specific whilst those of sustainability such as energy consumption or CO_2 emissions are common across several industry sectors.
- In setting aims and targets for indicators it is important that these do not contradict others. For example, the productivity indicator values employing minimum staff per unit of output – implying staff redundancies which would be in conflict with the aims of sustainability.

Synergies with UK water industry indicators

The water industry is an important market for MW and the company is looking to integrate indicators of sustainable construction with those prepared by the UK Water Industry to measure the specific impacts of these projects. Development and use of complementary indicator suites for specific industry sub-sectors is therefore important in addition to use of a standard group of generic indicators that apply to all work MW undertakes.

Ongoing process

Montgomery Watson is planning on implementing the sustainability indicators relevant to its business operations and is currently integrating sustainability issues with its Environmental Management System (EMS) so that this provides the data required. Early signs are that indicators on energy, waste and use of secondary materials will be implemented in 2001. Discussions have also commenced with major clients to investigate joint implementation of indicators to test on particular pilot projects.

This process, stimulated and encouraged by work with CIRIA, is therefore one of continual improvement.

CASE STUDY 4

Delivery of environmental improvements at the project level through environmental action plans

Improving the environmental performance of the construction industry requires delivery of high environmental performance at project level. The Environment Agency - as both a major government construction client and an environmental regulator - uses the Environmental Action Plan (EAP) within its management system in order to achieve high environmental standards on all construction projects undertaken.

The EAP summarises the protection, mitigation and enhancement measures which have been identified through the impact assessment stage and is reported as a chapter in the Environmental Statement or Report. The Plan is of particular importance in briefing those who carry the project forward to design, construction and management stages. It is written in such a way as to be readily adopted into contractual documentation as well use as a site reference document. A key feature of the EAP is that once agreed and published in the Environmental Statement the Plan is fixed and only modified following consultation.

Current Agency practice is to produce an Environmental Report or Statement at outline design stage. To be able to support this process, statutory and all other consultees must have confidence that appropriate environmental standards will be achieved throughout the subsequent project stages. These standards, which have been influenced by consultation and must be adhered to throughout design and construction phases, are set out as a series of targets and objectives in the Environmental Action Plan. The EAP is then adopted as the basis for the environmental brief to designers and the environmental specification to contractors.

High level objectives are set out in the Agency's Environmental Vision, for example, "Flood defences will be designed and constructed to deliver optimum environmental benefits". The EAP translates this vision into relevant environmental objectives and targets for the work linked to particular actions at project level.

The EAP also provides a record of all consultations relevant to the implementation of works. Mitigation of impacts and enhancement opportunities are detailed in landscape or ecological plans. The EAP also provides a valuable project management tool, ensuring that knowledge is explicit and always accessible and that sustainability issues are translated into clear objectives.

During construction – heavy steel piling forms a flood wall in front of existing structures.

The Environmental Action Plan ensured protection of the environment during construction as well as successful implementation of mitigation and enhancement measures.

After completion – the river frontage has been enhanced by the creation of a quality urban space and riverside walk.

Continued on next page

Illustration of typical targets within an EAP for an urban flood defence scheme

Objective	Target	Example actions
To protect all existing vegetation	No damage to trees. All tree works to be implemented in accordance with Arboricultural Association best practice	All trees to be protected by chestnut paling fence., 1 m outside the circumference of the leaf canopy. All fencing to be in place before each phase of the work commences
To ensure no pollution to land, air or water	No pollution incidents	Compliance with the Environment Agency's Pollution Prevention Guidelines
Aggregates used to include some recycled materials	10 per cent of total aggregates to be recycled	Surface all permanent access tracks with recycled stone
No damage to soil by compaction	No damage to soil structure by compaction	Existing track only will be used for access and storage. Tracks will be clearly marked and low ground pressure tyres used where appropriate
To enhance recreational value of the river to the community and access to riverside	Creation of riverside walk and adjacent sensory garden to the approval of the EIA Officer. No environmental incidents or public complaints	All hard and soft landscape works to be implemented and maintained according to contract, then signed off and handed over by the EIA Officer
To sustain collaborative and community initiatives related to the flood defence works	Enable and support the launch of the Wharf Road Primary School nature conservation pond within the site. Safe public access – no incident	Construct earthworks to satisfaction of contract administrator. Create public access on launch dat, providing a safe clean viewing platform for 50 people

NB It is usually helpful to indicate accountability for individual actions.

CASE STUDY 5

Company economic and social indicators for sustainability

In February 1999, WSP started a group-wide sustainability initiative leading to the development of company sustainability strategy to:

- work across the Group and extend our skills base, integrating the principles of sustainability in all that we do
- establish a dedicated Sustainability Consultancy with experts in economic, environmental and social sustainability.

The sustainability strategy launched in January 2000 set out a structured approach to achieve three key objectives:

- demonstrate leadership and set new standards for implementing sustainability for our clients and our business
- strategically extend our services to clients through applied research, innovative design and inter-disciplinary team working
- raise awareness, build understanding, develop skills and get buy-in from people throughout the Group.

These objectives are relevant to the company's three key stakeholder groups: shareholders, clients and staff. To facilitate meeting the objectives targets and indicators were established to measure progress. In the first phase of the strategy (2000) delivery of the objectives was process-orientated, reflecting the need to drive the initiative into the organisation. The indicators that were selected were therefore:

For leadership objectives:

- number of events, working committees and advisory groups participated in sustainable development for business and construction.

For raising awareness:

- number of sustainability roadshows delivered around the Group
- the production of key documents such as a sustainability brochure, guidance material, case studies and a dedicated intranet site.

Residents and designers discuss development plans in Charlotte Street

For services to clients:

- number of flagship projects won (integrating social elements of sustainable development).

Tangible measures of success for the business so far include:

- the establishment of a sustainability group with 12 dedicated staff
- an institutionally acceptable HQ building for WSP designed (in house) to sustainability standards
- reporting capability to Global Reporting (GRI) standards.

Successes in integrating social progress within business operations include:

- client and employee needs surveys, for sustainability services
- sponsoring the Engineers for Sustainability project, lead by Forum for the Future
- investment in training and long-term employment opportunities for people from Soweto, SA
- Community team building and capacity building in Africa and Asia.

Sustainability 2001 will set out the strategy for the next year. This will include indicators and targets based on our learning so far and chosen to enable WSP to manage, measure and report on the direct and indirect impacts of our business on people.

CASE STUDY 6

ICI's Challenge 2000 environmental objectives

In 1995, ICI published its Group 'Challenge 2000' objectives for improvements in environmental protection, health, safety and product stewardship to be achieved over the next five years.

The environmental objectives included percentage reduction targets for key waste releases from manufacturing sites. Categories included global warming, acid gases, hazardous emissions to air, VOCs, ozone depletors, aquatic oxygen demand, acids and ecotoxics to water, and energy.

In order to maximise the environmental benefit of investment in those businesses and sites where releases had the largest potential impacts a new concept of environmental burden was developed. Environmental burden is determined by multiplying the mass release of each waste product by a factor based on the substance's potential impact. For example, in the global warming category, carbon dioxide has a global warming potential (GWP) of 1, whereas trifluoromethane has a GWP of 11,700.

Progress and forward investment plans were reviewed during the annual budget review to ensure delivery of the objectives. During the period ICI's business portfolio changed rapidly as businesses were bought and sold. As a result the 1995 baseline had to be readjusted to remove the contribution from divested businesses and add the contribution from acquired ones. New businesses were required to adopt the Challenge 2000 objectives yet the progress made by divested business was removed back to the baseline year.

Progress towards objectives was published annually to demonstrate publicly ICI's steady improvement in safety, health and environmental performance. This improvement provides the baseline for Challenge 2005 – ICI's next set of improvement targets reflecting its new speciality portfolio.

CASE STUDY 7

John Laing Group sustainability report

John Laing Group is preparing a web based corporate sustainable development report for publication in Mid May 2001 (www.laing.com) in which examples of performance data and graphs can be viewed. The primary objective of the report is to provide clear, orderly information on the environmental, social and economic aspects of our activities, products and services. It is intended to enable more balanced decisions to be made about investments, procurement and partnerships.

The report follows the Global Reporting Initiative (GRI) format describing the company profile, policies, management systems and stakeholder relationships. A vision and strategy section is also developed. Most of the data within the report is at a corporate level for the year 2000. Where possible, data was collected for 1998 and 1999 to provide a reliable baseline on which to compare and interpret performance improvement.

Site visit by local schoolchildren

The sections on environmental, social and economic performance adopted the CIRIA strategic level indicators and tested these as they were developed. Based on stakeholder consultation, GRI general key performance indicators were also selected.

Where the management systems are robust, relevant data, for CIRIA, DETR and M4I operational level indicators was collected to meaure and report project site performance. Some of the CIRIA strategic indicators were not particularly relevant to John Laing Property and Investment operations within the Group. For these businesses chain control information and case studies were collated.

Laing Homes sponsored football team

Performance reporting has achieved clearer definition of the John Laing Group policy objectives and facilitated development for internal communication. Strengths and gaps have been identified within the Group management systems. The performance data will provide more focused strategic planning and future target setting. This will aid the drive to achieve continual improvement in environmental, community and economic performance.

CASE STUDY 8

Key performance indicators for energy use in construction

For office-based businesses, developing an indicator to monitor company energy use is relatively straightforward, as their premises are fixed and the majority of their machinery and plant operates from mains power. This means data can simply be collated from utility invoices. For the construction industry the challenges of compiling accurate data are much more demanding.

Carillion has been developing an energy KPI for the Group, to monitor progress towards reducing energy consumption. The work has identified a number of major issues for construction companies embarking upon this exercise:

- projects are of very variable size and duration
- only head office functions are fixed
- work is performed at a large number of sites that constantly change
- business structures may change making baseline data inconsistent
- projects are performed under joint ventures and with subcontractors
- a range of energy types and plant is used
- energy charges are sometimes included within office rental costs
- the types of projects in progress, and when these are undertaken, affect energy demand.

Having identified the logistical challenges of developing an energy KPI Carillion has identified a practical and pragmatic solution. Since all energy used on contracts is paid for, a system has been formulated using invoices submitted for payment. By using relevant cost and emissions factors to convert from the cost of fuel and electricity to energy usage a KPI of energy usage (obtained from bills) per person on site has been adopted.

Balfour Beatty

Sustainable construction indicators – development process

Balfour Beatty Major Projects maintains a suite of environmental performance indicators at both company and project level. To support the CIRIA RP609 project, a selection of indicatorswas trailled to assess project performance and ease of data collection. The trial was carried out at the Channel Tunnel Rail Link (CTRL440) Ashford to Folkestone project.

The project team also ran a separate workshop to identify additional indicators that could be developed to benchmark performance and drive the sustainable construction agenda.

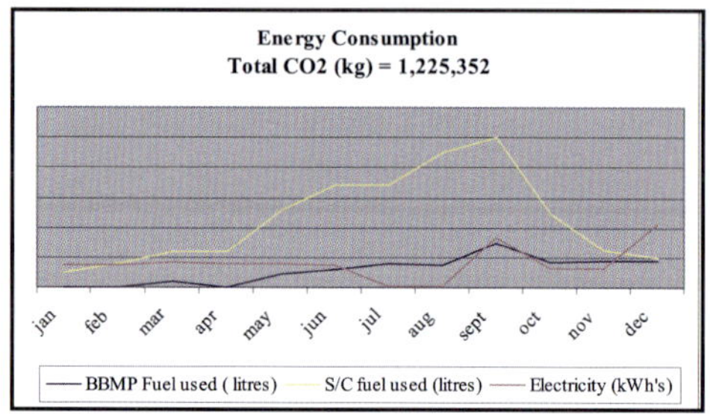

Use of energy

Energy consumption was measured from as purchased records. These included fuel for plant and equipment, and electricity used in the site offices. It was evident that the established data collection systems were designed to collect financial data and thus that usage data could not be readily extracted. This data therefore had to be collated manually – an onerous undertaking and one that demonstrated the value of integrating usage data into standard procedures.

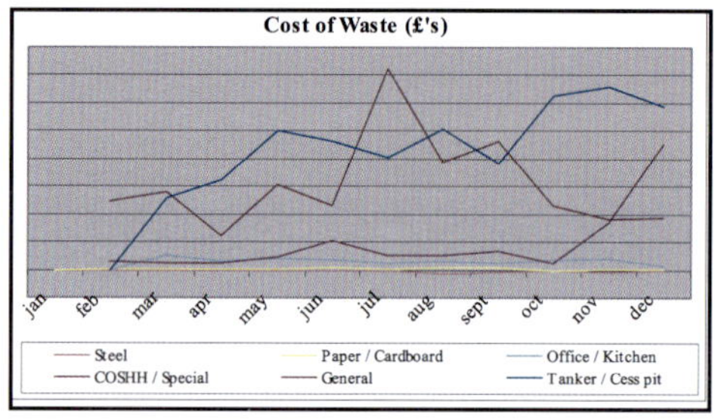

Waste removed from site

Waste removed from the compound was measured in the categories of special, general, steel, wood, paper and tanker waste (from office drainage). It was not possible to identify the split between waste sent for recycling and waste set to landfill.

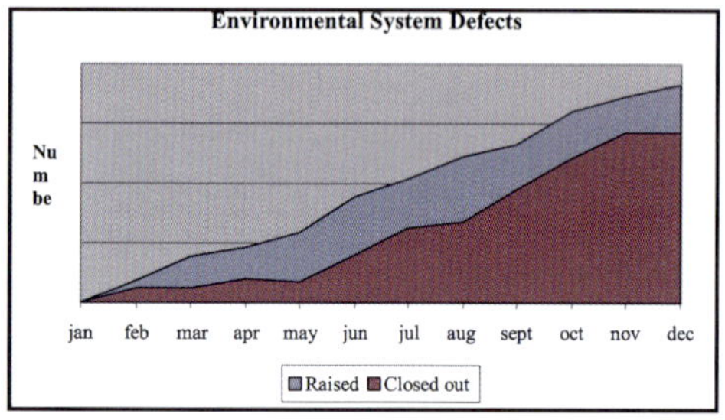

System defects analysis

All notifications for corrective action or non-compliance were recorded and presented graphically. This had the benefit of allowing trends to be analysed and, by including completion figures, showed how many actions were still outstanding. A powerful indicator to improve site standards.

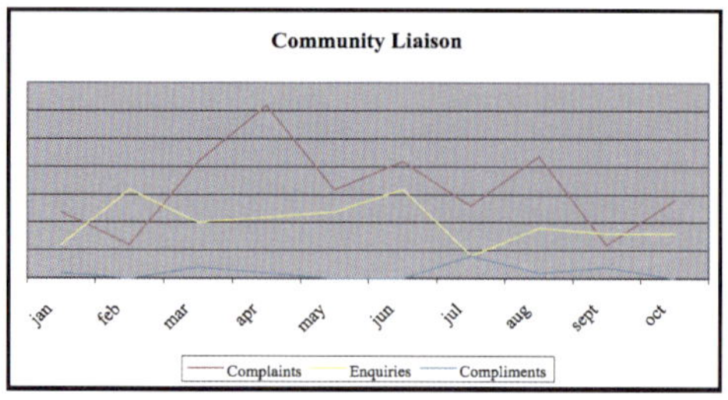

Complaints and community liaison

Complaints received were recorded and compared against site activities and community relation initiatives being undertaken. This proved to be a useful indicator to assess project specific performance, but was more difficult to use as a comparator between projects.